Ecology Out of Joint

By Lorus J. Milne and Margery Milne

—FOR THE GENERAL READER

Ecology Out of Joint
The Secret Life of Animals:
   Pioneering Discoveries in
   Animal Behavior (with
   Franklin Russell)
The Animal in Man
The Arena of Life: The Dynamics
   of Ecology
Invertebrates of North America
The Cougar Doesn't Live Here
   Any More: Does the World Still
   Have Room for Wildlife?
The Nature of Life: Earth,
   Plants, Animals, Man, and
   Their Effect on Each Other
North American Birds
The Ages of Life: A New Look at
   the Effects of Time on Mankind
   and Other Living Things
Patterns of Survival

Living Plants of the World
Water and Life
The Valley: Meadow, Grove and
   Stream
The Senses of Animals and Men
The Mountains (with the
   Editors of LIFE)
The Balance of Nature
The Lower Animals: Living
   Invertebrates of the World
   (with Ralph and Mildred
   Buchsbaum)
Plant Life
Animal Life
Paths Across the Earth
The World of Night
The Mating Instinct
The Biotic World and Man
A Multitude of Living Things

—ESPECIALLY FOR YOUNG PEOPLE

Because of a Flower
The How and Why of Growing
The Nature of Plants
When the Tide Goes Far Out
The Nature of Animals
The Phoenix Forest

Gift from the Sky
The Crab That Crawled Out of
   the Past
Because of a Tree
Famous Naturalists

Ecology
Out
of Joint

---

*NEW ENVIRONMENTS AND
WHY THEY HAPPEN*

Lorus J. Milne and
Margery Milne

Charles Scribner's Sons / NEW YORK

For helping round out our illustrations of major wildlife we particularly appreciate the assistance of Bob Hines and Beatrice Boone of the U.S. Fish and Wildlife Service, Department of the Interior.

Copyright © 1977 Lorus J. Milne and Margery Milne

Library of Congress Cataloging in Publication Data

Milne, Lorus Johnson, date.
　　Ecology out of joint.

　　Includes bibliographical references and index.
　　1. Ecology.　2. Animal introduction.　3. Plant introduction.
4. Wildlife conservation.　I. Milne, Margery Joan Greene, date.
joint author.　II. Title.
QH541.M523　　　574.5　　　76-48933
ISBN 0-684-14846-3

1 3 5 7 9 11 13 15 17 19　V/C　20 18 16 14 12 10 8 6 4 2

Printed in the United States of America

For Larry and Marilyn Staples
whose awareness and concern
range far from the practice of medicine
to the human environment, from New
England and Des Moines to all shores
of our world

Contents

The aim of science should be to take away the mystery from natural phenomena, but not to take away the wonder.

—*F. Fraser Darling*

Prologue

A NEIGHBOR and fellow naturalist calls himself a farmer or a juggler according to his mood. His two brothers say they are jugglers too, although one runs a small department store and the other hauls whole houses and barns from one place to another. "We all have to be jugglers," the farmer-brother insists. "A farm consists of countless parts, some going up, some coming down, just like a business or a huge load going along a byway. The boss's job is to juggle the parts and keep the enterprise in motion."

"A balancing act?"

"Sort of, except it can't stand still. It has to move along."

"The universe is change," one of us quotes from memory. "Nothing happens to anybody which he is not fitted by nature to bear," adds the other.

"Whose words are those?"

"The Roman emperor Marcus Aurelius, about eighteen centuries ago. Between battles, he kept a notebook which he called *Meditations*. It's amazing how much of what he wrote applies equally to the world today. Without scientific evidence to go on, he believed in nature as law-abiding. He wanted mankind to obey those natural laws while striving for self-sufficiency."

"Sounds like an environmentalist."

"A conservative, certainly," we agree, "yet aware of change. 'Everything is in a state of metamorphosis,' he wrote, and 'Forward, as occasion offers.' "

"Would that be ecology or evolution?"

"A bit of both. Perhaps human ecology in a changing world. Culture evolves so much faster than nonhuman life that the natural environment is pushed off balance. The ecology is out of joint in far more places today than during the times of Marcus Aurelius. Yet the cultural causes have not changed all that much."

Inevitable links between the past, the present, and the future lead us to continual exploration. Perspective helps broaden our intent. Life everywhere holds our attention. It excites us to sing out, to share personal discoveries.

We see the Mediterranean world of Marcus Aurelius through the eyes of George Perkins Marsh, a Vermonter who served for twenty years as U.S. minister in Italy. A scholar as well as a diplomat, he recognized the reduced state of nature in the former Roman Empire and feared that it could be duplicated in his native New England. To warn against this danger, he wrote *Man and Nature* (1864), a long-neglected classic of conservation and ecological thought. The focus of his concern shows in the subtitle for his book: *Man the Disturber of Nature's Harmonies.*

Profoundly alarmed by what he saw in southern Europe, Marsh sought a radical readjustment of civilization. Anything less seemed sure to continue the disruption of nature's balances until mankind and perhaps all life would become extinct. He regarded the progress of human culture as a "war against the order of nature" caused by ignorance and a loss of respect for the harmony to be found among nonhuman life. "Nature, red in tooth and claw" (as Alfred, Lord Tennyson, described it) appeared to embody constructive action to Marsh, for he believed predators to be essential in maintaining the proportions among the numbers of different species. Only domesticated animals, and then largely because of the conditions of their domestication, destroyed the sources of their own nourishment: the vegetation and then the land itself.

The examples to which Marsh pointed cannot be faulted. Yet he saw too that many instances of a worsening environment developed slowly and insidiously. No one person could be blamed for them. Effect shows no obvious link to cause or to means of correction if the change takes a long time. Only a government is likely to have the needed continuity and power to identify and repair damage to the environment. Marsh distinguished between a poor government and a great one according to the degree of their "atonement for our spendthrift waste." With governmental aid, man might "become a co-worker with nature in the reconstruction of the damaged fabric which the negligence or the wantonness of former lodgers has rendered untenable."

Efforts to restore the harmonies that Marsh praised are often postponed and minimized by governments because too few citizens subscribe to a belief that man's place on earth is as a tenant; man has the right of "usufruct but not . . . consumption," of stewardship and enjoyment of the harvest of each year but not of exploitation of its source. Even to sustain the annual yield to meet the needs of the human population in any given region becomes ever more challenging because of the accidental or unwise introduction of foreign plants and animals.

We like the way the distinguished English ecologist Charles S. Elton views these often-unexpected additions to the living community from elsewhere. He admits that

> invasions of animals and plants and their parasites—and *our* parasites—will continue as far as the next Millennium and probably for thousands of years beyond it. Every year will see some new development in this situation. That is a way of saying that *the balance between species* is going to keep changing in every country. Quarantine and the massive campaigns of eradication are ways of buying time—though they are valu-

able and necessary, they are also extremely expensive. It takes so few individuals to establish a population, and such a lot of work to eradicate them later on.

Elton's inclusion of parasites makes us especially sensitive to news about these inconspicuous invaders. Recently, in Nova Scotia, a friend asked us, "What's all this I hear about dogs in New England having heartworms? Is the infection spreading this way?" We knew some of the story, as all dog owners are likely to, but decided to explore the subject further by consulting a parasitologist.

"Heartworms are a new hazard in the American Northeast," the specialist informed us. "What's worse, they turn up occasionally now in house cats, in foxes, in coyotes, and even in people. Yet only fifteen years ago, heartworms were virtually unknown except from the Carolinas to Florida or near the Gulf of Mexico. A really alert veterinarian might diagnose heartworms in New England and immediately know that the dog had been taken far south during the warm months. In the Southeast, the infection is transferred by biting mosquitoes."

"Does this mean that we have to keep our dogs behind screens all summer?"

"The alternatives are not much better. You can give your pet a pill every day from the beginning of the mosquito season to the end, just as you yourself might take an antimalarial pill regularly in malarious country. The treatment is effective so long as too many infected mosquitoes do not get a chance to bite and overwhelm the chemical defense system. Malaria works the same way. Or you can watch for symptoms of an infection and then arrange for treatment. But it's drastic and dangerous, for the drug is potent. It may kill the dog, like an overdose of poison, or cause its death from the presence of so many dead worms inside its blood vessels. Heartworms by the dozen can grow in the pet's bloodstream. Their cylindrical bodies are nearly an eighth of an inch in diameter and six inches long when mature. They tend to lodge in the heart and render that organ too inefficient to keep the dog alive."

We realize that more people than ever before are like ourselves in driving the interstate highways to Florida for a holiday each winter. We can see their dogs enjoying the trip and know the dilemma of motel-owners in deciding whether to admit pets. More people, more pets, more decisions. And now more heartworms in places where no one previously detected the problem. All of the counter-measures appear impractical: to get rid of the biting mosquitoes that transfer the infection; to quarantine the pets that have been exposed in one part of the country before they are taken to another; to search out and destroy any wild reservoir in which the parasites might hide.

Efforts to shut out potential troublemakers of any kind are never completely effective. Yet they can reduce the numbers of newcomers for which the living community must make space. The goal is moderation in the rate of change, not a static environment. Progress is important, but only at a pace that does not trip up the living world.

Fresh concerns command our attention so insistently that we often overlook the rate at which new dangers replace old ones. How little we are plagued today with Old World insects that the early colonists from Europe introduced: the head louse, the human flea, or the clothes moth. The cabbage butterfly and its hungry caterpillar, like the housefly, are still with us but not in such numbers as to cut deeply into the food supply or to transmit diseases in a major way. We have learned to live with the codling moth, introduced to America in 1750. Yet in 1909, it was causing more than $12 million damage to fruit crops, despite over $4 million expended to control it. The San José scale insect, which came from eastern Asia in 1870, is now widespread but of minor significance. The Mediterranean flour moth, which came to the United States in 1889, has been shut out by better packaging. The shiny scarab known as the Japanese beetle is no longer the scourge it was soon after 1916, when it became established in New Jersey around Riverton. But the gypsy moth from Europe (1868) continues to damage important trees. The cotton-boll weevil, which was carelessly allowed to spread north from Mexico in 1892, circumvents every attempt to control it. So do the fungal diseases that devastate American chestnut trees (from Asia in 1904), white pines (from Europe in 1906), and American elm trees (also from Europe in 1930).

Quick action, despite immediate costs, becomes necessary to exterminate each new infestation before it expands. A program of this kind, paid for in the late 1930s by the Rockefeller Foundation, rid the New World of an African mosquito that turned up in northeastern Brazil. A carrier of malaria, the mosquitoes reproduced in sunny pools far from forest shade and entered houses to bite people more readily than the native mosquitoes. Three years of intense effort and $2 million brought success. More recently, Florida citrus growers spent a larger sum to meet a different challenge when it threatened to destroy their industry. Mediterranean fruitflies appeared in a few groves, and fly maggots began burrowing into the fruit, rendering it unsalable. Concerted efforts wiped out the invaders. A new infestation appeared in 1975, this time in California. A different counterattack was mounted in just a few weeks, using a technique devised for a different insect (the tropical screwworm fly). A special cargo of male Mediterranean fruitflies was airlifted from Hawaii and freed in the infested orchards. These insects had been raised in captivity, and each one sterilized by high-energy radiation. Just as with the screwworm flies, which had begun a serious affliction of cattle, the female fruitflies mated with the treated males and then laid infertile eggs. New shipments of sterile males arrived every week or ten days, insuring that each fresh generation of impregnable females of the species in California would be promptly serviced. Gradually the number of fertile eggs would reach zero, and the battle end. Admittedly the technique is costly. But it affects only the target species and does not pollute the environment.

Native pests require vigilance too if introduced crop plants are to thrive. True, no one today fears the Colorado potato beetle. The tomato hornworm, which used to cause so much damage, is scarce. Now educational supply companies raise these spectacular caterpillars in captivity for students to examine. We see a similarity with the effort necessary in urban medical schools to find sufferers from famous old-style diseases so that interns can learn to recognize the symptoms. In developed countries, a person rarely has a deadly dysentery, diphtheria, typhoid or paratyphoid fever, scarlet fever, or malaria. Yet these diseases were lethal everywhere at the beginning of the century. They cannot be forgotten, although other challenges now hold our attention. Effective measures are still being sought for infestations of the native caterpillar known variously as the corn earworm, the cotton bollworm, the tobacco budworm, the tomato fruitworm, or the false budworm, according to how the insect is encountered in its versatile diet.

Even the depredations of native insects on native plants of economic value plague us. The spruce budworm, which used to ravage northern forests only about twice each century, now strips millions of acres of trees that have been earmarked for pulpwood. Thereby it affects paper companies, employment, tourism, and taxes. The white-pine weevil continues to destroy the callow "candle" on many a pine tree, requiring its growth skyward to follow a side branch as a new leader. This deforms the trunk, often several times in the long life of a pine. Lumbermen deplore a deformed pine today just as much as did the officials of Queen Anne when they attempted to enforce her 1711 "Act for the Preservation of White and other Pine-Trees growing in Her Majesties Colonies . . . for the Masting Her Majesties Navy." The weevil persists, despite plenty of attention, although no "Surveyor General of Her Majesties Woods" penalizes any person who dares "Cut, Fell or Destroy any White or other sort of Pine-Tree fit for Masts."

Understandably, insects interfere with many human plans. Of all the species in the animal kingdom, each one with its own way of life, insects account for fully three-quarters. Some of these small, fast-reproducing, moderately adaptable animals can make destructive use of almost everything that people cherish. Yet introduced insects have brought gains as well as losses. Australians virtually rid their continent of prickly-pear cactus, which escaped from cultivation and spread too far, by introducing a tiny cactus-miner moth from Argentina in 1926. Ranchlands in the American West are almost free of the poisonous "Klamath weed"—a European Saint-John's-wort—because of the activity of some small shiny beetles brought from South America.

The costs of disharmonies in nature grow greater. The economic benefits wrought by the establishment of new harmonies or the restoration of old ones grow too. So do the satisfactions in undoing past mistakes and minimizing new ones. Yet an informed appreciation of the present shows us that it is always too soon to relax, to say Amen. Life offers no finality. Our goal must be progress toward a balance, however precarious.

That goal is within our grasp, although seizing upon it may entail a good deal of discomfort. Much depends upon a growing adjustment in cultural values. Fortunately, more people in more places than ever before realize the interdependence of all life. Earth's resources are shared resources, and the opportunities of each individual diminish as the human population increases. With the earth's population already beyond 4 billion, up from about 1.6 billion in the year 1901, we are well along toward a second doubling in the twentieth century. Many human ecologists hope that without worldwide calamity this second doubling in human numbers will be the last.

That the members of any species should elect to limit their own reproduction is something new on earth. Species other than mankind have evolved strategies with the same effect, but only as a consequence of the combined pressures from the living and nonliving environment. We can voluntarily join the club—actually rejoin it, after failing to pay our dues for nearly twenty thousand years.

So major a revolution in human behavior, and the recent

events to which it can be attributed, merit the fullest comprehension. We know that, given time, life can adjust to enormous change. Some kinds of animals and plants will lose out locally or become extinct. Other kinds will replace them, perhaps diversifying under a new set of environmental conditions. But what happens when the natural rate of change is greatly speeded up through human intervention, whether deliberate or inadvertent? The gradual evolution of a new ecologic order is tumbled into near chaos. The outcome grows less predictable. This book explores what changes are natural and what effects (both good and bad) the spread of civilized mankind is having on the world.

An Ecological Conscience

"A thing is right when it tends to preserve the integrity, stability, and beauty of the biotic community. It is wrong when it tends otherwise."

—Aldo Leopold, "The Land Ethic," in *A Sand County Almanac* (1949)

"Wilderness values . . . are, indeed, so basic to our national well-being that they must be honored by any free society that respects diversity. We deal not with transitory matters but with the very earth itself. We who come this way are merely short-term tenants. Our power in wilderness terms is only the power to destroy, not to create. Those who oppose wilderness values today may have sons and daughters who will honor wilderness values tomorrow. Our responsibility as life tenants is to make certain that there are wilderness values to honor after we have gone."

—William O. Douglas, *A Wilderness Bill of Rights* (1965)

1

Natural Diversity

H A V E   Y O U   E V E R   R I S E N at sunrise on an extended trip and asked yourself where you were? In a first-class hotel room the answer might not be evident. Perhaps you have only to look at an electric wrist watch and read the day and date, to know what place you must be according to a re-membered schedule. "If this is Tuesday, it must be Belgium!"

Suppose in your travels you took a nap on a deserted bit of sandy beach. Awakening, could you tell by looking around you where on earth it is? How and in what ways would an Atlantic coast differ from a Pacific, or even an ocean margin rather than that of a big salt lake or sea?

Were you to stop your rented car at the side of a road in the midst of a grassy plain to take a rest, might you not, upon waking, ask yourself if this is a prairie province of Canada, or Australia, or Argentina? The make of the hired automobile—a product of civilization—might give you the answer more quickly than any of the natural features of the landscape.

Every year this little game of "Where am I?" becomes more challenging. No longer can you notice a few white egrets associating with a group of cattle and conclude safely, "I must be in the Old World, somewhere between Portugal and the Caucasus, or southward in Africa to the latitude of Ghana to Mozambique." Those particular birds might be in Florida now, or Puerto Rico, or New Zealand. Since the 1930s they have spread spectacularly and adopted new lands in all directions from their former home. In only one place have we found them failing routinely: on the Dry Tortugas Islands seventy-five miles southwest of Key West, Florida. Lack of fresh water and of food let dozens of egrets die, too weak to try elsewhere.

We used to believe that a few palm trees growing out of doors, rooted in the soil, would tell us we were close to the tropics, if not actually in the frost-free Torrid Zone. Southern Florida, with its native palms, offers this subtropical ambience. So does Palm Springs, California, where a visitor may wander amid tall *Washingtonia* palm groves in adjacent canyons. But then palms thrive close to the Irish Sea on the west coast of northern Scotland, at the latitude of central Labrador. Horticulturalists at Poolewe Gardens started these admirable plants where the warmth of the Gulf Stream would reach them. Many other areas of our planet are equally suitable for palm trees. No longer can we rely upon any single type of life to indicate our approximate location on earth. Too many areas have now been altered by mankind.

The contours of the land and the climate change too without human intervention, and sometimes faster than is appreciated. Six thousand years ago, the Ice Age had not quite ended in northern Scotland. Great Britain was no "tight little island," but an oddly shaped peninsula projecting northward from France. To reach England, a person, a wild animal, or a plant had no water barrier to cross. Even the highest tides could not flood into the present location of the English Channel. The quaint Channel Islands were just high points in a lowland plain. The local climate was too cool to encourage colonists of many kinds. And by the time the weather improved, the land level had changed.

The Channel opened. Now a British naturalist can tell with ease which shore is which merely by the absence or presence of specific plants and animals. The French coast (but not the English) has asps and Aesculapian snakes, terrapins, and a dozen (not just five) kinds of frogs and toads.

History records the confusion caused by confrontation with large animals from remote lands. Alexander the Great profited in this way when he brought back from conquests in India a few regiments of well-trained war elephants. Apparently Ptolemy, his general in Egypt, inherited these elephants when Alexander's empire was divided. Was this the source of the thirty-eight elephants with which that wily leader, Hannibal of Carthage, started out from his base in Spain on a roundabout route through the Italian Alps toward Rome? Some Carthaginian coins of the day show the small-eared Indian elephants, fitted out for battle. Other coins depict large-eared African elephants similarly equipped. Perhaps we will never know whether Ptolemy supplied these impressive beasts as an early version of lendlease. Hannibal could have obtained African elephants almost as easily, for herds of them were still common in the foothills of the Atlas Mountains. Has anyone trained those giant beasts in the Indian way?

By the time of the Crusades (twelve centuries after Hannibal), not an elephant survived north of the great deserts in Africa. Gone, too, were the Asiatic lions—the lions of Xerxes and of King David of Judah. The only remnant survives on the forested Gir Peninsula of western India. At present, a pride of lions in open country is a mark of African highlands, from Ethiopia southward to Kruger National Park.

Gradually one learns to recognize the living signs of a region. A mob of kangaroos or a forest of eucalyptus trees identifies Australia. Any hummingbird indicates a New World location. More specifically, a ruby-throated hummer will be encountered from spring until autumn only east of the Great Plains in the United States and Canada. Each island of the West Indies seems to have its distinctive hummers, such as the magnificent streamer-tail of Jamaica or the diminutive bee hummingbird of Cuba.

Contrasts with previous experience alert each of us. This became evident recently when we had the pleasure of making a field trip with two visitors from abroad through a woodland in New Hampshire. The naturalist from England and the biology teacher from New Zealand both found chipmunks especially intriguing, for these common and trusting squirrels with stripes had no counterparts in their home lands. The Englishman knew gray squirrels, for they have been added to the British scene.

"The grays have about done in our European red squirrels, you know," he commented. "The reds are a shade larger, but not nearly so aggressive. To see reds abundant now, you have to go to the Isle of Wight, where they have no grays for competition."

The New Zealander found all squirrels exotic. To him a woodland with chipmunks, gray squirrels, and American red squirrels seemed all but unbelievable. After all, New Zealand has no native climbing mammals of any kind. There even the brush-tailed possum, which was introduced from Australia in the hope of starting a fur trade, is rarely seen. These cat-sized marsupials are nocturnal and secretive. Although locally abundant and quite destructive of forest trees, they rarely encounter people, or vice versa. No doubt this explains why the only possum we met in New Zealand, on a night walk through a public park, made no attempt to escape. We brought our flashlamp so close that the glass protecting the bulb actually touched the animal's pink nose. The possum was not tame, merely unused to people. Outdoors anywhere in darkness, people tend to be rare.

Occasionally, while conferring with someone who has not yet fully adjusted to the environment around a new home, we ask, "Is there something here you miss, compared to what you remember about your previous location?" One

response made by a New Englander who had moved to
California comes to mind: no fireflies. That the one state
with luminous millipedes lacked light-producing insects
had not occurred to us. We could recall any number of
places in the tropics and the Temperate Zone where fireflies winked their communication signals in the dark. Each
species, we knew, has a coded message of its own, letting
males find females of their particular kind.

To discover luminous insects that are native to lands with
humid nights is to gain a new appreciation of the differences among living things that await a traveler. Fireflies
and glowworms, which are their young, have no monopoly
on light production. A recent experience makes us realize
how often a substitute may be on hand and, in itself, a distinctive feature of the place. We were strolling through the
fragrant darkness of a warm night along a narrow road beneath subtropic trees in southern Florida. Suddenly, we noticed beside our path two pairs of small greenish yellow
lights twirling slowly, about three feet above the ground.
Detouring to investigate, we discovered that two click beetles with luminous spots were entangled in a spider's web.
The spider was investigating too, although we could not see
her until we used a flashlamp. She ignored the light and
concentrated on her task. First she cut enough of the
strands in her orb web that she could spin the beetles faster
while she immobilized them with a swaddling of silk. Then
she drew them together and wrapped the two in a single
package. All the while, the beetles' glow shone through.
Soon it would end. The energy in the beetles would become energy in a spider and then energy in a spider's eggs.
The spider was harvesting a surplus of click beetles and
producing a surplus of little spiders. They would nourish
birds and lizards, recycling the nourishment in the same
region.

Our memories drifted back to other places where we had
encountered counterparts of these luminous click beetles,
other insects that had a bright spot above and below at each

side of the body. The particular insects had been "fire beetles" in southwest Texas and much larger *cucujos* in Panama. Just one of these amazing creatures, when pressed against newsprint in complete darkness, gives enough light to let a person make out the words. A dozen in a jar have sufficed to allow a surgeon to perform an emergency operation when all other sources of illumination failed.

Ordinarily these differences among states and countries are sustained over a very long time. Only one or two conspicuous changes can be expected to occur naturally in any one part of the world during a human lifespan. Ernst Mayr of Harvard University tells us that the ancestry of all the land birds native to the Hawaiian Islands can be traced back to about fourteen kinds of birds that reached these remote landfalls. They arrived at an average rate of one new kind every twenty-five hundred years. Early arrivals found a need to share the islands mostly with migratory transients. This left numerous opportunities for survival in different ways. Gradually those ancestors diversified as they exploited the various ecological situations. Mayr believes he can recognize the sequence in which the ancestors came, from the degrees of uniqueness shown by Hawaii's breeding birds. Earliest of these settlers would be those that gave rise to the distinctive honeycreepers. Most recent would be the black-crowned night heron, for it remains identical with its North and South American counterparts.

Without human intervention, the rate at which the newcomers replace the old residents is slow. Inconspicuously, a few species lose their ability to compete. Their geographic range shrinks. Finally, no more individuals survive to reproduce their kind, and a species disappears forever. A different species fills the gap. So long as these changes occur just two or three times in a century, nature appears in balance.

The balance of nature is, of course, an illusion. Each community of wild plants and animals seethes with quiet actions and reactions. An inventory at year's end is like a count of the ballots after an election. Which members of the community are still in power? Which have been subdued but might recoup their losses and influence the future? Have any disappeared or withdrawn into some refuge as though it were a fortress?

To pretend that any place on earth is free of human influence is an illusion too. Every place in which people can scratch out a living has already been occupied. Today the change consists in replacing small family groups and tribes at the subsistence level with large commercial enterprises. Each enterprise may prosper with a minimum of human supervision if it produces something that can be sold elsewhere; it must pay for the energy required to extract the product from the land and get it to market. A salable crop becomes the goal. Rarely is it any native form of life, from which an observant person might learn the answer to the question Where am I?

Often we can only guess, as we look at a field of cotton, what natural vegetation and associated animals would live on the area if the climate and the minerals in the deep soil were the only controls—if nature could only struggle undisturbed. Great territories would revert, we know, to forests of slow-growing trees and abound with forest animals. Other areas, deprived of irrigation, would become hospitable again solely to the life forms of semiarid lands. These plants and animals have benefited from millennia of slow evolution by accumulating a genetic heritage of special adaptive features. These fit them to survive summer drought and winter cold.

Yet many regions that formerly contributed their fair share toward incorporating energy from the sun into food for life may never regain their wealth. The wetlands that have been filled—particularly the salt marshes along the seacoasts—have no way to renew their uneven contours or again to welcome earlier tenants. Until recently the oceans, which are so vast that they could easily swallow up the continents and still be more than eight thousand feet deep, seemed forgiving of all the wastes dumped into them. Now man-made organic compounds are poisoning the green plants, and jetsam from civilization is fouling the shores and bottom.

A truly enormous diversity of animals and plants has been replaced for human benefit over vast areas of the world. Occasionally the change has been conspicuous because large animals were supplanted by introduced kinds on a continental scale. Nineteenth-century America witnessed such a conversion as millions of Old World cattle and sheep took the place of millions of native bison. A complex food network, of which a few Indians had been the principal human beneficiaries and which no one had yet analyzed, was exchanged for a short food chain: prairie soil to grass, to domesticated livestock, to people of European origin. But the old balance of nature did not convert easily into a new equilibrium, even though the relationships within the short food chain were so obvious. The use of favored living species to make energy and special substances available for human populations in America led rapidly to depletion. Charles S. Elton focused on the most flagrant feature and identified it as a repeated error: "Overgrazing or mismanaged grazing, and soil erosion has often completed a process that may end with the shortest food-chain of all—nothing."

Today, much of the former bison territory serves to produce grains and plant fibers for mankind. Yet each well-kept field in which such crops are raised contains far fewer kinds of life than a barely productive desert. A man-made forest ("tree farm") of favored species is virtually a desert too. The diversity and abundance of native life in any of these agricultural areas is as close to zero as the custodians can keep it.

Dreams of taking the land into partnership to obtain a living still appeal to many young people. They see idle acres and get permission to prepare the soil, put in favored plants, work to protect the crops, and use the harvest.

These teen-aged or slightly older amateur farmers reject the structured life-styles of their parents as too dependent upon petroleum resources and too polluting. Instead, the young people take pride in using manpower, not machines and pesticides. But experience proves a hard teacher and soon reveals why previous farmers abandoned the land. In good years it may support a subsistence enterprise but yield no surplus of sufficient worth to meet the demands of inevitable harsh seasons. Nor can it sustain the cost of indispensable services of civilization.

Replacement agriculture may be economical where the climate is naturally suited to particular crops. In these limited areas, it is easier to improve toward higher yield both the crops and the croplands presently in production than to convert land elsewhere and maintain it at a gainful level. Unprofitable regions are left to subsistence farmers, whose success or failure matters mostly to themselves. Rarely can such marginal people protect the soil from erosion while baring it for crops and keeping it exposed to the sun. Inevitably, the wealth of life shrinks. Extinctions multiply. The Law of Conservation of Matter and Energy still applies. But there is no corresponding law of the conservation of species.

Where the climate is benign, ever more energy must be turned toward preventing or suppressing outbreaks of pests in the deliberately simplified communities of life. Some of these destructive agents move readily from wild vegetation to introduced plants in productive areas. Far more become pests after arriving from elsewhere on earth. Conscientious citizens apply helpful techniques of quarantine, but the system resembles a sieve with many holes. Whenever a potential pest gets through, only swift, intensive action at high cost has a chance to eradicate it. If not detected in time or treated with insufficient effort, the new population is likely to explode. Even when later controlled, it must be tolerated too as a permanent resident. "Control" reduces the annual losses but extends this rate into an indefinite future. As Canadian orchardist A. D. Pickett observes, "We move from crisis to crisis, merely trading one problem for another."

Simplification for efficiency proves expensive. Yet civilization continues to depend upon live plants and animals because it cannot synthesize food at any reasonable cost. Technology has replaced the carrier pigeon, vegetable and animal sources of dyestuffs, the silkworm, and the physical power of man and beast. Yet these advances offer a false hope for self-sufficiency, while temporarily avoiding the complications in managing a live community outdoors. The substitutes for plants and animals are machines and their products, "fed" with coal and petroleum from the geological bank. Now that the scientific monitors of the bank are warning of an imminent closing of these accounts, the wisdom of diminishing withdrawals becomes more widely appreciated.

It seems ironic that the dire predictions of the Reverend Thomas R. Malthus were ridiculed until the present century, ever since 1791, when he first published them anonymously. The ecological disaster he forecast for mankind failed to occur by the time the human population reached one billion. How could he have guessed that agriculturalists would find convertible land in the American West, in the Canadian prairies, in Argentina, and in Australia? How could he imagine that research in plant genetics and animal breeding would so spectacularly increase the yield per acre? The grim spectacle of two-thirds of mankind malnourished or starving did not develop for almost two hundred years. By then the number of people totaled four billion, and the finite world was identified as the lonely "Spaceship Earth."

Today the human lives, the costs, the populations of rare and endangered native plants and animals, the extinct species, and the uncontrollable outbreaks of diseases and pests are all counted. Reluctantly, since our own kind is so implicated, we can identify the causes. Meriting our own self-given name of *Homo sapiens*—the wise species of mankind—

is not so easy. Ignorance and greed have already cost too much. What can be done to stabilize the situation and then progress at a reasoned pace? Where are we all, and where is there to go?

2

New Arrivals

SUNDAY AND MONDAY, March 2 and 3, 1975, afforded "the birding event of the century." Excited people in heavy clothing congregated to glimpse a rare bird along the north shore of the Merrimac River where it opens to the sea near Salisbury, Massachusetts. With luck, they focused their high-powered binoculars, spotting scopes, and expensive telephoto lenses on one gray-winged gull scarcely bigger than a city pigeon. Its squat, dovelike proportions, short legs and neck, and wedge-shaped tail all made it unique.

Experts from as far south as New Jersey responded to telephone calls by rushing to the discovery area. Each confirmed the identity of the mystery bird: a Ross's gull from northeastern Siberia. Never before had one been seen on an Atlantic coast. The only earlier recorded sightings in America were a few in October, seaward from Point Barrow, Alaska, and from Canadian points along the Arctic Ocean.

So improbable a rarity on the Massachusetts coast required positive recognition by several birders: Paul Miliotis of Dunstable, Massachusetts, Edward and Martha Gruson of Concord, Massachusetts, and Walter Ellison of

An improbable rarity on the Massachusetts coast! Ross's gull near Newburyport, on March 6, 1975. Its short beak, wedge-shaped tail, red feet, and pigeonlike flight were distinguishing features for the bird since, although adult, it showed neither the rosy color nor the narrow black neckband of mating season. (Photo by Richard A. Forster, Massachusetts Audubon Society)

White River Junction, Vermont. They enthusiastically shared their news, almost like Paul Revere. Only then did less confident observers admit that they had seen the stranger as far back as January but had been unable to give it a name.

So far as is known, Ross's gulls are scarce even in their limited home area. There they alternate between brief summers on or near their boggy breeding grounds and long winters ranging far from land over the frozen ocean. Wherever winds and currents shift the ice floes and expose open water, these gulls pluck the larger plankton and small fishes from the surface. This habit brought the birds to the attention of the British explorer James Clark Ross in June 1823. He and another member of the expedition each shot and preserved a specimen near Melville Island, close to the North Magnetic Pole, and carried them back for presentation to scientists in Great Britain. The Scottish ornithologist William MacGillivray gave the species its scientific name and referred to it as "Ross's rosy gull" because of its unusual coloration. "Cuneate-tailed gull" was offered subsequently as a common name, referring to the tail shape, which is evident all year, not just in the breeding season.

The sight of so remote a gull rewarded the world-famous bird artist and field man Roger Tory Peterson. To see it for himself, he left his Connecticut home at 3:34 in the morning. Ross's gull was to be the six hundred and sixty-eighth kind of bird on his Life List for the United States and an item to include in the next edition of his popular field guides. The *New York Times* carried an account of the bird on page one, completing the story inside. Soon readers all along the Atlantic seaboard knew what to look for: the peculiarly wedge-shaped tail, red legs and feet, and a tinge of pink on the breast feathers below the folded wings. The pink was easier to imagine at sunset, for it does not develop fully until early summer.

A wide audience watched the gull on television, captured by color cameras for a television news program on the following Friday evening. Some Chicago people, wintering in warm Miami, decided to wing home ahead of schedule, detouring through Boston to the chilly site of the small gull. Soon after, *Time* magazine accorded more than half of its first page to an illustrated account of the "visitation" and implied that "the descent of a Martian spaceship" would scarcely have been more astonishing. A man and his son from Windsor, Ontario, tossed sleeping bags and winter camping equipment into the family microbus and drove by night to reach the Massachusetts coast. They arrived before dawn, took a nap, and awoke to see the ultraspecial bird only a short distance from their parking place.

Experts quickly noted that the Ross's gull was an adult. A juvenile would have been far more likely to venture the thousands of miles beyond the normal range for its kind. Usually these young "erratics" travel with regular migrants of another species. Possibly the rare visitor from Siberia made a similar mistake, despite its maturity. As though suffering from a bird's equivalent of an identity crisis, the Ross's gull kept company with a small flock of Bonaparte's gulls. These have a similar size and scavenge on the mud flats when not nesting. Their breeding grounds are in the far Northwest and easternmost Siberia. After their young can fly, Bonaparte's gulls migrate diagonally across North America to winter along the southeastern shores. Some of them regain their black head feathers and begin courtship on the Gulf coast of Florida, before starting back toward the arctic tundras. Would the lost Ross's gull follow its adopted companions to Siberia?

As we stumbled across the coastal grasslands to join the throng of people observing Ross's gull, the diversity of occupations represented and the camaraderie born of this shared interest warmed us. Nothing else did! A cabinet officer—the secretary of defense—was there with us, on a side trip from his stop at Pease Air Force Base in nearby New Hampshire. We were all shoulder to shoulder with lobstermen, filling-station attendants, grocers, university professors, off-duty policemen, a banker, and curious retirees. Some of these people filled their hours of waiting for the feathered will-o'-the-wisp to return from wherever it went at high tide, by watching other erratics. A tufted duck from northern Europe had settled in a tidal creek less than a mile away. Four separate snowy owls from the Arctic surveyed feeding territories on the extensive salt marshes within five miles to the north.

We waited too and wished that some enterprising lunchwagon salesman would arrive with a truckful of hot coffee. A Ross's gull, a tufted duck from Iceland, or a snowy owl must be better clad than we to withstand March winds, with their chill factor augmenting the cold of clear air from Canada. We thought of the studies that have been made recently on the homing ability of birds and wondered if these erratics had what they needed to find their way back to their ancestral breeding grounds. Ross's gull gave no answer. It hung around Newburyport until mid-May. Christopher W. Leahy of the Massachusetts Audubon Society described the bird as "languishing among the Bonaparte's and ring-billed gulls like an aging starlet, her public dwindling." The gull's breast feathers took on an "unearthly" pink, and it gained a narrow black collar. Where is the bird now, and did it ever find a mate?

We think of earlier expeditions to see birds that had come from far away. Each time it seems fair to question whether the first one or few betoken colonization to come. Could an inherited way of life be adjusted enough to succeed in a new environment? This measure of versatility must be hidden among the bird's genes at least as deeply as the determination of the shape of its tail or the hue of its mating plumage.

Years ago, a special summons hurried us to hear the forceful "cheer, cheer" call of a male cardinal, singing in a suburb of Toronto, Canada. None of his kind had ever been seen there before. One of our companions wondered whether it had been a caged bird, repeating an old practice that has been illegal for many years. Had the cardinal escaped far from home and potential mates?

That cardinal may have been the first, but others followed. "Redbirds" now nest over much of southern Ontario. They have expanded their range far beyond the southeastern United States. Their colorful plumage, cheerful song, and generally beneficial eating habits are all appreciated. Illinois, Indiana, Kentucky, North Carolina, Ohio, Virginia, and West Virginia all claim the cardinal as the state bird.

What resources must a region offer to induce cardinals to stay and raise a family? We see these birds mostly where a residential area affords an array of fruit-bearing shrubs

and trees. Cardinals in winter relish the sunflower seeds
and bits of unsalted peanuts that people put out for birds.
As a consequence, northern members of this species are
mostly city-dwellers, even more so than robins. Yet the contents of the crops of cardinals that have met accidental
deaths reveal that wild seeds, wild fruits, and cultivated
grains rank high in the diet. About two-thirds of the food a
cardinal eats each year is plant material. The rest is chiefly
beetles, true bugs, caterpillars, and grasshoppers. The proportion changes according to season and latitude—more insects in spring and in the southern parts of the birds'
range.

Cardinals and wild grape vines so regularly act as
partners that it is easy to predict the territory of cardinals
by noting every reachable area where wild grapes grow. No
matter how shriveled the grapes may be in summer, fall, or
winter, a cardinal will hunt them out. In the southern
prairies, these birds manage to find leftover grapes even in
springtime. Later the cardinals drop the grape seeds where
they have a chance to sprout. But the recent expansion of
cardinal territory seems the result of the looser partnership
existing between these handsome birds and landscape gar-
deners, who recommend the planting of particular shrubs
and low trees. Among such vegetation the cardinals find
nesting sites and food, including insects that otherwise
would harm the woody plants. Usually the male finds a
high lookout point at which to sing while he watches for
any need to defend his nest area.

The sunflower seeds on a birdfeeder in New England
today attract far larger numbers of another thick-billed
bird that formerly was unheard of—the evening grosbeak.
We recall responding to an invitation to observe a small
flock of these noisy birds in New Jersey, where they had
suddenly appeared. Not since the winter of 1889–90 had a
vagabonding group of this kind been recorded so far east.
Evening grosbeaks have not moved out of their native
Northwest, but they now arrive each winter on schedule in

New England and adjacent Canada, southward into Georgia. They gobble down any cedar berries the waxwings have left, then fight for handouts at feeding stations. Even the bullying bluejays find urgent business elsewhere as soon as the evening grosbeaks move into the shrubbery.

Whenever we learn of a new arrival, we wonder if the event should tell us something. Is it a portent of more to come? Has the environment itself changed? And if potential colonists are coming now, what kept them from moving this way a thousand years ago? Or did they move, perhaps many times at long intervals, but fail to establish themselves? After all, it was not Leif Ericson in A.D. 1000 or Christopher Columbus on any of his four voyages to the New World who succeeded in starting the European invasion of America.

To this degree, history repeats itself and provides a basis for judgments about the future. Since natural historians first began to observe the continent, several kinds of mammals have become familiar in states and provinces of North America where they were not known a few decades before. In each instance, these animals have been able on their own to reach and occupy territories that were formerly closed to them. The difference in their fortunes began when people from the developed countries of Europe made major changes in the environments of the New World. This is of small comfort to the independent truck farmer in Florida who has to compete with burrow-making armadillos for marketable crops. Or for the Maine poultry farmer whose loosely penned flock is raided successively by the same lone coyote and then by an opossum that gains entry to the henhouse. Why did not all of these animals stay in the Southwest or, better still, in Mexico?

The opossum was already well established in coastal North Carolina when Captain John Smith and his party arrived from England in 1607. Indians showed Smith how easy it was to kill an opossum with a stick, how good the flesh was when cooked, and how useful its shaggy fur could

Growing and fattening on a wide variety of foods and hiding during the day wherever shelter can be found, opossums expand their geographic range without much effect on the environment. Only about thirteen newborn bee-sized babies in each litter of twenty can survive their first day by finding and swallowing a nipple in their mother's pouch. When they leave the pouch fewer manage to cling piggyback and accompany the parent as she forages, yet enough mature to replace the old and supply colonists. (Photo by Frank M. Blake, U.S. Fish and Wildlife Service)

be in making garments to keep human bodies warm despite the winter chill. Their Algonquin term for a white animal, *apasum,* became *opossum* in English, often abbreviated to *possum.* At that date the creature was probably common all through the southern United States, suitable areas of Mexico and Central America, and southward to Argentina, as it is today. Smith recorded that "an opassom hath a head like a Swine, a taile like a Rat, and is of the bigness of a Cat. Under the belly she hath a bagge wherein she lodgeth, carrieth, and sucketh her young." He did not mention that truly white individuals are rare. More usually, the black or cinnamon-brown underfur shows through the sparse white-tipped guard hairs, leaving white the neck and three lengthwise strips on the head. This animal was the first marsupial mammal of any kind encountered by Europeans. It is also the largest marsupial in the New World and the only one that has yet extended its range beyond southern Texas.

The felling of the forest east of the Mississippi River opened up the country and offered the opossum more of the type of habitat it had already exploited in the southeastern states. Primarily the animal is a scavenger, a nocturnal opportunist, able to find nourishment and shelter almost anywhere. Hollow logs and piles of brush that people tolerate at the edge of a woodland suit the opossum particularly well because they are close to open country in which to search for food. Such places are scanned by day by human neighbors, who unwittingly protect the opossums by getting rid of native predators. The role of bobcats, weasels, large owls, and similar animal-eaters is rarely taken by domestic dogs and cats. These household pets do not get hungry enough to devour an opossum. Its habit of feigning death upon discovery—and frequently of emitting a foul odor from secretions near the anus—deter predatory behavior in the dog or cat and send it elsewhere to practice pouncing on something more responsive.

After the time of Smith, the opossum expanded its range northward about two miles per year, until it encountered long cold winters. Since the animal is not a hibernator, further progress beyond southern Vermont and New Hampshire seems unlikely. After hiding from a storm for a few days and nights the opossum must come forth for food even if the temperature is low. Its ears, its tail, and sometimes its feet get frozen and later erode away. Without the help of its prehensile tail, the opossum has difficulty climbing, gathering material with which to insulate a nest, or steadying its young while they ride piggyback on forays in the dark.

No one who has had experience with opossums gives them credit for more than a minimum of intelligence. Even one of our friends who raised one after another of these animals as house pets has few compliments for the creature other than that it almost always made use of a box of "kitty litter." Often the opossum snarled at her approach and bit her if she picked it up. Every one of its fifty white teeth is a sharp cone capable of penetrating human skin. Its mouth, in fact, has more teeth then any other kind of land mammal and closes with more muscular force than might be expected. Sometimes our friend's current "pet" would feign death at her approach, causing her to complain that the opossum was treating her like a dog! Nor is an opossum more sociable with others of its kind except during the brief breeding season, which occurs twice annually in warm regions and once in cooler habitats. After permitting copulation, the female rejects the male and may drive him from the forty acres or so that she accepts as home.

Maternal behavior in these marsupials combines a strange mixture of seeming solicitude and total disregard for the young. Within two weeks after mating, the mother bears as many as twenty young, each the size of a honeybee. On their own, they must creep into her pouch. There she has nipples for only about thirteen of them. The first to find and claim this food supply have a future; the rest of the litter perishes, ignored by the mother. Within two weeks the survivors will be as big as mice. At one month of age, the young can free themselves from the nipples and begin to peer out at the world from the security of the pouch. Their mother will keep them with her for two months more, then send them off on their own.

The taste of opossum meat, which Southerners like best combined with sweet potatoes, has led Frank B. Clark of Clanton, Alabama, to establish both the Big C Possum Ranch (which doubles as a drive-in theater) and the Possum Growers and Breeders Association, Inc., of which he is president. He sponsors an annual International Possum

Show, at which animals with superior efficiency in meat production are auctioned off as breeding stock. Clark's enthusiasm over a stud male weighing thirty-five pounds—three to four times as much as most wild females and five to six times the weight of the usual male—may prove contagious. Opossum-raising could spring up far beyond the present range of the animal in the wild. Already this versatile marsupial has become established in California, Oregon, and Washington. Some are managing to perpetuate themselves in thornscrub country, which verges on desert. Where they will spread from there is less an idle guess than a vague dread for people who raise the many crops that opossums raid at night.

The wild coyote excites much more comment as it extends its range. Urban dwellers become uneasy when they discover that these wily doglike predators are around. Many farmers and hunters regard any species of carnivore as a most unwelcome neighbor. What warning should be recognized in the coyotes' calls? Certainly they communicate to distant coyotes regarding food, perhaps mating, and danger. What will a coyote attack? Obviously neither children nor adults are threatened unless a coyote is rabid. Often a guard dog responds to a coyote as though it were just another dog—and the coyote may play this game too. Whether the outcome may be a "coy-dog" is still moot. Separating fact from myth is not easy. Generally, however, the coyote relies upon its individual prowess. It hunts alone, with a mate nearby, or accompanied by a few half-grown young. Its favored foods are rodents, which a single coyote stalks efficiently. Seldom does a coyote assail any prey that will struggle vigorously or fight back. Instead, it takes what comes easily, especially carrion. At a large carcass, such as that of a deer, it will tolerate a dozen companions, each gorging on its own. Fruits, including fallen apples, comprise fully 15 percent of the coyote's diet.

The Spanish conquistadores in Mexico were the first Europeans to meet a coyote or hear its song. They adopted the Nahuatl word, *coyotl,* for the animal, merely changing the last letter to make the word easier to pronounce. By 1847, English-speaking settlers in the Southwest came to know the *cayeute,* or they pronounced *coyote* this way without changing the spelling. The early pioneers, whatever their language, regarded the animal as a small version of the wolf. Often coyotes accompanied packs of wolves on the Plains or hung around Indian encampments, devouring refuse without creating any conflict. An Indian accepted the coyotes' nocturnal music, even incorporating an imitation into his own tribal chants. He regarded the living creature as the embodiment of a major spirit in his animalistic religion. No other wild neighbor seemed as likely to have been the assistant of the Supreme Creator during the days when the living world and man were formed.

Europeans changed the world of the coyote. First, they introduced cattle and horses west of the Mississippi and killed many of the progeny solely for tallow and hides, leaving carrion from the slaughter to scavengers. Soon the pioneers discarded the bodies of bison in immense numbers, removing only the salable pelts for robes and occasionally the tongues as delicacies. Wherever this unsought food became plentiful, coyotes multiplied and extended their range. When people traveled overland to Alaska, along the way killing horses that faltered and foals that could not keep up, the coyotes followed and took up a place where they had held none before.

Soon herders with flocks of sheep arrived to claim western lands. The stock animals they brought came from a long line that had been bred for centuries to be docile and to convert anything green into mutton and wool. Sheep could not defend themselves from an aggressive coyote. Often a sheepherder would find himself equally helpless, although armed with a gun. The coyote could simply run swiftly to the opposite side of his flock and make a kill, while the sheep themselves blocked the herder's fire!

The noted observer of life in the American West J. Frank

Ordinarily both coyotes of a mated pair hunt for food they can bring home to their pups in the den. This parent has not yet interested the one youngster to emerge in the half-grown jackrabbit brought home for breakfast. Research reveals that rabbits and rodents, insects, and other small animals constitute about 60 percent of a coyote's diet; fruits and other plant materials, 15 percent; and carrion, the rest—if it can be found. (Photo by E. R. Kalmbach, U.S. Fish and Wildlife Service)

Dobie, after nearly a lifetime in Texas, found himself on the side of the coyote. At least this predator afforded a delaying action on the conversion of sheep country into desert. Dobie's position appears clearly in his book *The Voice of the Coyote:*

> Coyotes are the arch-predators upon sheep. Sheep are the arch-predators upon the soil of arid and semi-arid ranges. Wherever they are concentrated on ranges without sufficient moisture to maintain a turf under their deep-biting teeth and cutting hoofs, they destroy the plant life. . . . Metaphorically, the sheep of the West eat up not only all animals that prey upon them—coyotes, wildcats and eagles especially—but badgers, skunks, foxes, ringtails and others.

Sixty years of systematic destruction of coyotes and other predators have not greatly changed this situation in the West. The native animal is still regarded as a "varmint," to be poisoned, trapped, or lured by tape-recorded calls of natural prey (rabbits, not sheep!) within easy range of a hunter with a telescopic sight on his high-powered rifle. Hunters and sheep ranchers credit the coyote with neither a prior right to follow an ancient way of life nor a beneficial role in holding down the number of rabbits, rodents, and insects. These people ignore the sheep-made deserts of the Old World. They see no reason to admit that their introduced sheep, rather than the coyotes, keep the ecology out of joint.

Any coyote that moves northward or eastward out of sheep country escapes persecution in the West. The resourceful animal even has ready-made travel routes remote from human habitations. It can trot along below the high-voltage power lines of utility companies, where mice and other food are fairly plentiful. The power companies apply selective herbicides along the path of their lines to prevent woody growth and thereby provide corridors of imitation prairie over hill and dale. The prairie warbler too has come east, apparently along these grassy avenues.

Probably the first coyote to reach Maine, Ontario, or New Brunswick had to wait a while for a potential mate to catch up. Occasionally the delay was shortened, because people adopted coyote pups as pets in the Southwest, only to release them in the Northeast as young adults at full vigor. Now officials of fish-and-game commissions in New England are learning how to distinguish a coyote carcass from that of a feral dog and the differences in behavior of the live animals.

Today, when the moon brightens the rural Northeast, it is often possible to see a coyote or to hear a group of them raise their voices in quavering chorus. A naturalist who experiences such a meeting through eyes or ears feels fortunate to have been present at the right place and moment.

Our most recent reward of this kind came in January 1975, close to the Canadian border in upper New York State. The animal's sound always reminds us of nights spent camping in warm Arizona or New Mexico, years before the coyotes spread so far east. Or we think of western coyotes that survive in California, despite perils of so many kinds. Millennia ago, a few of these predator-scavengers got mired in the tar pools of La Brea, where Los Angeles now maintains a historic park. Coyote skeletons, virtually complete, have been recovered from the tar, along with bones of saber-toothed cats, dire wolves, mastodons, and other long-extinct beasts.

Before the arrival of people many a coyote escaped to maintain the species, after feasting on these ancient carcasses. Live coyotes still find food and refuge in hilly parts of Pasadena, just a few miles from the tar pools. These persistent westerners tune up on moonlit nights, just hours after the same clear skies have stimulated a similar response among coyotes in the East. Now coast to coast, as well as from Costa Rica to Alaska, these intelligent animals are holding their own. They will likely serenade in their wild way for millennia to come.

So far as we can detect, nowhere has a crisis for any native animal or plant arisen from the arrival of coyotes, opossums, evening grosbeaks, or cardinals far from their former range. The same might be said for cattle egrets and several other kinds of birds that have crossed barriers to establish themselves in distant lands. The animals themselves managed to arrive and settle in with a minimum of assistance or none at all.

The unplanned role of mankind in providing a habitat for new colonists is evident today in the extreme north. Increasingly in the last few decades, discarded oil drums and equipment rust in the summer sun across the arctic tundra, where the low vegetation scarcely grows from year to year and no trees rise to conceal the human litter. Bush pilots refer to the empty drums as "flowers of the Arctic" and see them everywhere that civilized man has been. Yet each drum with an open bung hole offers an opportunity to a pair of snow buntings. The sparrow-sized birds seek out these havens every June, often before the ground is bare. The would-be parents make countless trips with strands of moss and bits of grass or sedge to build a nest in semidarkness. They line their product with wisps of fur and feathers. The sheltered space under the mudguard of any abandoned car will serve the buntings almost as well. Somehow they manage to get their fledglings out of each hideaway.

Snow buntings have always nested farther north than any other land bird. Refuges of man's making were all they needed to extend their range across flat tundra to the shores of the polar ocean. Where few members of this species visited two decades ago, flocks of them now congregate like house sparrows. Until they must hunt midges and other small insects to feed their young, the buntings congregate. Flocks of them rise in unison, undulating in flight or veering in sudden curves like magic carpets of feathery snowflakes. Each bird exposes its distinctive white wing patches and a clean white underside. All the while they whistle and trill like "the laughter of children," as John Burroughs describes it.

The impact of these new arrivals on their adopted environment seems as gentle as that of a butterfly settling on a flower. In the modern world, even a butterfly can find fresh opportunities. The monarch (or "King Billy") of southern Canada and much of the United States is now admired elsewhere as the "wanderer." To start a new colony it needs only some milkweed plants in a part of the world where none grow naturally. Sooner or later a pregnant female monarch arrives. She finds the plant and lays her eggs on it. Caterpillars hatch out, each one banded narrowly with encircling stripes of yellow, black, and white. The larvae grow and transform into butterflies of a new generation. Rarely do they damage the milkweed significantly. It produces its full quota of flowers and attracts honeybees as well as local insects to nectar and pollen. Later its green pods ripen and split open, freeing glossy white silken hairs that catch the breeze and pull out one flat brown seed after another. The hairs support it, airborne, to travel and start new milkweeds, often miles away.

Milkweeds from North and South America have crossed the broad Pacific Ocean, mostly as seeds caught on cargo. Occasionally a thrifty person has given the weed some help. One gendarme in Quebec did so soon after he accepted a new position in the South Pacific. When he was told to take along whatever he would need for comfort, he suspected that a soft bed might not be available to him in far-off New Caledonia. He looked around the autumn landscape and began collecting milkweed fluff from ripe pods. He stuffed a whole bag full of this material and took it with him on his journey. After the matted fibers had served their purpose in a sleeping pad, he discarded them—along with some seeds that were still alive. The seeds became milkweed plants. Somehow, monarch butterflies found the weeds on the remote islands.

Milkweeds and monarchs were together in Hawaii even earlier—in 1845—but no one knows how either kind of life reached those islands. The butterflies established an outpost in the Marquesas Islands in 1860, New Caledonia and the east coast of Australia in 1870, and New Zealand four years later. Today the wanderer is common at many places in the East Indies, on the Malay Peninsula, and in Burma. Its caterpillars thrive wherever their food plant is well established.

The first butterfly of this kind recorded on the eastern side of the Atlantic Ocean turned up in England in 1876. By 1962, no fewer than 214 more were caught and reported in the British Isles—as many as 30 in some years. Yet not a single one of these insects was seen en route. Many butterfly fanciers imagined that a strong wind kept the fliers high above the water and propelled them eastward, perhaps at an average speed of fifty knots. But a crossing from America would take at least fifty hours at this rate. Could those hours be consecutive? Would the butterfly fly in darkness? Surely it could not stand the chill if blown to great heights. Might it settle for the night on a wind-tossed sea and then take off again at dawn? A more plausible explanation would be that the insect settled accidentally on an eastbound ship, stayed motionless for a few days because of cold and wind, and then flew again when British shores were close. This idea strains credulity too, for no one has noticed a butterfly as a stowaway on any transatlantic passage.

Until 1962, even a pregnant monarch that reached the United Kingdom had almost no chance to produce a new generation. Milkweeds are not native to any part of Eurasia. The deficiency ended, however, when Brian Gardiner of Cambridge imported some milkweed seeds and cared for them in a special flower bed. When no monarchs arrived to take advantage of his hospitality, this butterfly fancier arranged to have a few mailed to him, each packed flat in a triangular envelope. Every one of these tough insects survived the trip. Said Gardiner, "After a good feed, they got on with the business of mating and were no trouble at all." Today, anyone who sees a monarch flying in Britain has no reason to dream of the butterfly's aerial odyssey from America. It may have come from less than a mile away. It can find milkweeds in flower gardens and waysides, and make itself at home.

In North America, monarchs put real effort into their traveling. They also demonstrate an amazing ability to find their way on an inherited schedule. Those butterflies that mature early in the year in the Mexican highlands or in the southern United States head north. They mate as they go. The females continue onward, laying their eggs on milkweeds wherever summer weather has just arrived. A few weeks later, a new generation of monarchs emerge and again wing northward. But by September, when the generation in the northern states and Canada becomes airborne, these fliers take a different direction—south.

These facts are known through the persistent studies of a biology professor at Scarborough College, University of Toronto. Frederick A. Urquhart and his helpful wife, Norah, work chiefly close to their home near Toronto, on the north shore of Lake Ontario. We assisted these dedicated people one late September, following them before dawn to a grove of tall oak trees along the lake shore. Hundreds of monarchs still slept in the morning coolness, clinging to oak foliage within reach of a large net on a long handle. Fred wielded the net. Norah and we transferred the butterflies he caught into a huge carton, the man-made darkness of which would keep the insects quiet. Later the morning's catch was carried to the Urquhart home and released inside a big screen porch. Now the task of tagging began, with four pairs of hands at work. Pick a butterfly off the screen and carefully rub the scales from a small area above and below the right front wing just behind its leading edge. Now curve a self-adhesive label, printed on waterproof paper, around the lead edge and squeeze it firmly together so that it will hold to the bared wing membrane. Let the butterfly go, and watch it as a flight test to be sure the paper tag has no effect on its normal locomotion. If all is well, get the next untagged butterfly and repeat the process. Not until all had been marked were they released through an opened door, a dozen or more at a time. Up they zoomed into the sunlit air, in single file, and headed south across the lake.

Monarchs with tags on which were printed "Return to Museum, Toronto, Canada" and a serial number were recovered near the Gulf of Mexico, some fifteen hundred miles from the Canadian sites where the identifiers were attached. Thousands more, from Alaska and western Canada, spend the cold months at Pacific Grove, on the Monterey Peninsula of California. There a local ordinance decrees a $500 fine for anyone caught disturbing the butterfly visitors.

Until December 1974, the occasional report of great flocks of monarchs southbound in Mexico remained something of a puzzle. The one clear instance of so long a journey was a specimen tagged near Toronto on September 18, 1957, caught again the following January 25 in San Luis Potosí; the straight-line distance between these points is 1,870 miles. What kind of a tail wind could propel the insect unharmed so far so quickly? Or did it actually fly an average of 14.5 miles daily for the 129 days, despite stops to sleep and sip nectar? Then came the surprise: a moving mass of live monarchs was discovered coating more than ten thousand trees and the ground between them, to a depth of eight inches on some Mexican volcanic mountains. Among them were monarchs tagged in the United States, which convinced the explorers that here they had found the main wintering site of this amazing insect.

Monarchs and their gaily striped caterpillars have intrigued naturalists for years because birds that eat insects so obviously avoid this particular kind. A young bluejay with no experience will accept either the butterfly or its larva as an edible morsel, swallow it, then react to the chemicals the insect body contains. Usually the reaction is quick: an upchuck. But the memory lingers, and that jay will refuse a second caterpillar of the same kind or any butterfly body that has the distinctively colored wings attached. Recent research shows that the caterpillars retain the poisons from milkweed "milk," to which they themselves are immune. In fact, a monarch caterpillar will drink this toxic liquid if it is offered and will show no ill effects. It retains the poison even into adulthood and thereby gains a way to teach a hungry bird or reptile not to sample a second insect of this kind. Relative freedom from attack permits most monarchs to flit where they choose and their caterpillars to feed in plain sight. Only a virus disease keeps the numbers of monarchs fluctuating and thereby saves most milkweeds from being devoured.

The arrival of a monarch creates no need for fear. Nor does the appearance of a milkweed plant far from its home in the New World (or in Africa south of the great deserts) constitute a threat to local life forms. Even domesticated livestock, out of which most kinds of wariness have been bred, find milkweeds distasteful. Only if these animals are seriously malnourished, on ranges depleted by overpopulation or ruined by prolonged drought, will they eat a toxic dose. Often the livestock that eat milkweeds under these circumstances are already only a few days away from death by starvation or thirst. The toxins in the milkweed merely hasten their demise. It is then too late to blame the milkweed for being on the range or to notice the more fundamental situation: all nonpoisonous plants have been totally devoured or withered without rain.

So often an obvious change is overlooked for quite a while. Or a casual observer feels amazed when a predictable sequel strikes home. Even the confirmation of the Ross's gull on the coast of Massachusetts gained impetus. Two birders from nearby Manchester admitted that they had seen the bird almost two months earlier from the Newburyport side of the river. Their record books showed this tentative identification. But when friends to whom they mentioned their suspicions tried to find the gull and verify its species, the bird eluded them. The earlier sightings were dismissed as highly unlikely. Probably other people had their binoculars focused on the same mixed flock during the succeeding six weeks but saw only Bonaparte's gulls because no others were expected.

Everyone, it seems, is conditioned by early training not to make extravagant claims. But perhaps the old story of the boy who cried "Wolf!" needs updating. Did the boy actually see a wolf the first time, only to have it run away before anyone could confirm his sighting? The moral from this tale may be that reporting a fact before others are ready to receive it often gets nowhere.

Any new arrival is worth watching. So is the community of life that must eventually react if the immigrant becomes a colonist and establishes itself. Nature works with a system of checks and balances so intricately adjusted that it often defies complete analysis, even with the aid of the most sophisticated computers. A gain in one direction brings a loss in another. The local plants and animals will accommodate quite well to an addition if they have time—time measured in decades or centuries rather than in some span briefer than a human generation. This is the pace that was normal for more than a billion years, before impatient mankind arrived upon the scene. During that vast time, life encountered numberless opportunities to refine its organization. It is unlikely to be disrupted by anything short of a massive change or the impacts in quick succession of one new arrival after another.

3

Cast-off Pets

FLORIDA'S FRESHWATER FISHERMEN rarely tire
of telling the true story of how walking catfish were discovered near Boca Raton. The historic event came on May 25,
1967, during a sudden squall, which knocked out the electric service to a construction site. The night watchman on
duty, W. A. Turner of West Palm Beach, ruefully set down
the story he was reading and reached for the flashlight he
kept ready above the work table. Just then, outside the
shack, his dog began to bark excitedly. Turner opened the
door and shone the beam toward the animal. That was
when he saw a pale wriggling body, which held the dog's attention. It looked like a white fish. Did fishes really ever
come down in a rain storm? There was no waterway within
half a mile.

Turner stayed in his doorway, holding the light beam
steady on the queer fish and his dog. To his amazement,
the creature headed straight for him. It repelled every effort of the dog to seize it but twisted itself as though sculling with its tail and supported its body on its pectoral fins.
When the fish got within range, Turner kicked it away. But
it turned and again came toward his light. This time
Turner caught the fish in a carton, where he could exam-

ine it more closely. The barbels on its lower jaw were clearly those of a catfish. But a *pink* catfish fully ten inches long with pink eyes that reflected his lamp? The thing must be an albino. Turner decided to keep it to show someone, for his dog was no help as a witness.

The next day, Turner got in touch with a newsman. The reporter took one look and telephoned to Vernon E. Ogil-vie, a biologist with the Florida Game and Freshwater Fish Commission. Ogilvie heads the Nonnative Fish Research Project for the whole state. He too hurried to Turner's home and inspected the captive. Ogilvie recognized it at once: an albino variety of the walking catfish *Clarias ba-trachus*, a native to Southeast Asia and eastern India. He knew that young fish of this kind had been imported for the ornamental aquarium fish trade in Florida ever since a German dealer described the exotic in a trade journal back in 1961. "I didn't know they walked," said Ogilvie. "You can keep the catfish, if you want it. Or kill it and eat it. Just be sure you don't let it loose!"

This was the first *Clarias* catfish that Ogilvie had seen alive after a capture in the wild in Florida. A few others had been caught, killed, and brought to him. Yet his concern over the fate of this particular fish proved redundant. Before mid-August, more than twenty-five hundred of these albino catfish were caught on roads or in ditches and ponds in the northern part of Broward County and adjacent parts of Palm Beach County. Almost certainly they were breeding. Probably they originated from Penagra Aquariums, a large fish farm west of Deerfield Beach. Others subsequently appeared in Hillsborough County, presumably after escaping from other fish farms in that region. Soon fish farmers too began to have problems with walking catfish, for these strangely amphibious creatures could invade a fish pond from a drainage ditch just as easily as they could go the other way.

At first, almost every walking catfish seemed to be an albino. Then dark gray or black ones were found. By 1972, the albinos had almost disappeared, but the dark individuals had taken their places and extended the range of the species. Perhaps fish-eating birds or native fish that could prey on young walking catfish selectively removed the pale-bodied ones and let darker individuals survive. Certainly they took advantage of the network of freshwater canals in southern Florida and spread into, and then around, Lake Okeechobee. They also invaded the Intracoastal Waterway, despite a salinity of about 1.8 percent—half that of full-strength seawater.

In late spring and early summer, these catfish begin mass migrations by both water and land as they search for a new habitat in which to spawn. By late summer, their young are two inches long, and by winter, as much as five inches or more. At any age they are voracious opportunists. In a small pond, one walking catfish will kill and eat most of the other animals. If two of the catfish are present, one may become a cannibal, yet soon be ready for another hearty meal. The catfish devour small fishes, particularly killifish, mosquito fish, and sunfish wherever these are numerous. Examination of the stomach contents of the catfish reveals the abundant remains of water beetles, immature dragon-flies, and mayflies, plus plant materials too. On their Florida diet, catfish attain eighteen inches in length, which is four inches less than the record from Thailand. Yet in an aquarium, this newcomer can also fast for up to eight months, losing weight only slowly and retaining fully its ability to become instantly aggressive if disturbed.

Like almost any catfish, *Clarias* has a stiff sharp spine at the leading edge of each pectoral fin. The spines bear a mucous coating and can inflict a painful wound. But the family specialty is a treelike breathing organ in each gill chamber, occupying a peculiar pocket attached to the second and fourth gill arches. So long as the fish can gulp air and keep these pockets filled, it has no real need for gills. In fact, *Clarias* will drown if prevented from reaching the air above its aquatic habitat. Muddy swamp water with virtually no oxygen is neither worse nor better than the clearest, most oxygen-rich water for this creature.

Ichthyologists are aware that the walking catfish has had man's help in becoming a breeding resident of both Guam and Hawaii. It has been released in both southern California and Nevada, with no known success in either place. Now several of the western states, including these two plus Arizona, Colorado, New Mexico, Utah, and Wyoming, have linked their wildlife agencies into a wildlife council to ban importation into, and transportation or possession within, their boundaries of walking catfish and sixteen other foreign fishes, as well as some reptiles and amphibians. For Florida, such legislation appears already too late. At least two dozen freshwater aquarium fishes are now represented by breeding populations. Almost as many more, including the dreaded piranha of South American rivers, have been caught in canals and ponds. The walking catfish is probably the least manageable of these introductions, because if the waters containing it are poisoned, it simply leaves the water and goes elsewhere. It can jump over a three-foot curbing as easily as out of a six-inch pail. Nor does it hesitate to dine on frogs, snails, dry dog food, fruit, or cheese if it encounters these on land.

Many Floridians fear that the walking catfish will devour so many of the smaller fishes and dragonflies that mosquitoes will soon become a serious pest again. If the newcomer gets into the waters of the Everglades, it seems likely to reduce seriously the numbers of the apple snail on which the endangered Everglades kite depends almost completely. Nor are anglers joyful over having a new fish to catch, one that challenges human abilities to design lures and then puts up a respectable fight. What do you do with a walking catfish? It dies slowly and chews up everything else in the creel. And although those who have sampled its cooked flesh attest to good flavor and texture, the job of removing the skin from a walking catfish is truly formidable.

So far, the amphibious fish has served mostly to focus attention on the impact of the pet trade and of cast-off pets on the living world, where these animals are foreign. A team of biologists at Florida Atlantic University in Boca Raton has gone so far as to indict these nonindigenous species as biological pollutants, since they degrade the quality of the environment. Those introduced forms that achieve dominance do so at the expense of native life and thereby reduce the all-important diversity of the community.

The open waters of southern Florida offer special opportunities for exotic fishes. An interconnecting network of irrigation and flood-control canals, ditches, and slow streams join hundreds of miles of man-made waterways for small boats. Warm rains and mild climate combine with well water at nearly constant temperature in sustaining this habitat through most seasons. More than 250 fish farms take advantage of these natural amenities, supplying nearly 80 percent of the total demand for aquarium fishes in the United States. To serve this market, about 110 million live fishes are imported annually, almost half of this traffic through the ports of Tampa and Miami. Much of the traffic through other ports finds its way to Florida fish farms to be held in ponds, cultured, or bred in captivity. Few precautions are taken to prevent deliberate or accidental release of ornamental aquarium fishes into Florida waters when tanks or ponds must be drained for cleaning or redesign. In addition, many a winter visitor to the state conscientiously conveys the living contents of the aquaria in his apartment or condominium to the nearest canal as a humanitarian act, year after year, before closing up for a summer absence.

Tropical-fish fanciers recognize most of the exotic fishes that have been caught recently in Florida waters:

Climbing perch
Siamese fighting fish
Dwarf gourami
Twospot climbing
 perch
Kissing gourami
Paradise fish
Pearl gourami
Blue gourami
Pacu
Caribe
White piranha
Blue acara
Oscar

Black acara
Rio Grande perch
Firemouth cichlid
Jack Dempsey
Pearl cichlid
Jewel fish
Blue tilapia
Mozambique tilapia
Walking catfish
Tinfoil barb
Rosy barb
Zebra danio
Goldfish

Carp
Armored catfish
Pike killifish
Broadspot molly
Swordtail molly
Guppy
Liberty molly
Black molly
Lyretail black molly
Green swordtail
Southern platyfish
Variable platyfish
Shortblade swordtail

(Various hybrids have also been found.) Only the tilapias and the carp seem to have been experimental introductions based upon the hope of gaining a new source of human food. All the others are past or potential pets.

Old-timers can still recall when nothing exotic could be found in local waters, and the choice of a pet fish was quite limited. The inhabitant of the home aquarium might be a goldfish, either plain or fancy, or some wild denizen from a brook or pond, if it would survive confinement in a bowl. We ourselves kept a three-spined stickleback for many months, probably until its average lifespan of 1.5 years came to a normal end. We enjoyed the company of a lonely pumpkinseed sunfish, which in a pond may live to be 2 years old. The domesticated goldfish may survive for 15 years, as a fair average between the 10 years of a housecat and the 20 of an old dog.

Then came the affluent custom of heating an aquarium and aerating and filtering the water; aquariums were illuminated so that plants would grow well. The favorite fish became the guppy, named for the Reverend Robert J. L. Guppy, who discovered these glittery little denizens of pools on the island of Trinidad. Guppies are native to the southern Antilles and eastern South America as far as southern Brazil. They are less belligerent than the closely related mosquito fish of fresh waters from New Jersey to northeastern Mexico. Both guppies and mosquito fish are viviparous, not oviparous. And although a guppy may live only a year, any aquarium is big enough to house a pair. The broods of young to which the female will give birth may not survive long, for the parents are both prone to cannibalism.

Fish fanciers rarely regard mosquito fish as suitable pets, perhaps for the same reason that few people would raise dandelions as house plants. Mosquito fish hold importance only in eating the wrigglers and pupae of mosquitoes that otherwise might bite people. No one could object to letting the mosquito fish perform this natural role in distant places. Introductions were made in the irrigation ditches of Arizona and California, and in the runoff from warm springs as far north as Alberta, Canada. To this latter area, a team of government fisheries officers went recently to check on mosquito fish that had been freed in 1924. The new study in the national park at Banff revealed tropical immigrants as well as mosquito fish in the out-flow water from Cave and Basin Hotsprings. Local aquarists had "planted" guppies, green swordtails, zebra cichlids, and sailfin mollies. All of these exotics darted about, completely at home. The brilliant guppies outnumbered the slightly larger mosquito fish and the others too. These displaced denizens of warmer climates survive the northern winters by courtesy of earth's inner heat, brought to the surface in the thermal waters.

So far as we can see, the presence of alien fish in local streams and ponds excites few children into collecting and caring for these prizes, as a modern counterpart to our own fascination with sticklebacks and pumpkinseeds. Possibly the extra glitter and the foreign origin cannot compensate for the special investment in a warmed aquarium to meet the needs of tropical swimmers. It is much easier to tend a wounded bird or an apparently homeless dog or cat. Their helplessness plays upon the empathy of a child, who will readily minister to the creature's needs. Smaller local animals can be maintained in makeshift quarters more adequately than those from distant places. This may account for the appeal of some long-lived crustaceans and insects, spiders and snails, amphibians and reptiles. No better way to get intimately acquainted with wild creatures occurs to us than to live close to them for months or years. Even while captive in an imitation of their natural habitat, they reveal an individuality we might not otherwise believe possible. To us, this is the real magic in ministering to a pet.

We have shared special corners in our home with ghost crabs. They taught us what to listen for when one ghost stridulates to another in the darkness on a sandy beach. Two tree crickets of the same species and unlike sex will shyly demonstrate their mating behavior while confined in a large terrarium in the house. He trills his claim to territory while emitting a subtle fragrance to lure the female. She creeps close enough to nibble on an attractive gland exposed while his vibrating wings are raised. He takes advantage of her distraction but keeps trilling until he has insured his fatherhood of her still-unlaid eggs.

In season, a frog or toad will summon others of his kind to an orgy, regardless of his location at the water's edge, indoors, or close by in the garden. A tree frog chirps at night when the relative humidity is right, even though the month for reproduction is long past or still to come. He signals that he is here! We can enjoy the sound, without wishing to isolate the frog from his fellows or his normal diet. When in Jamaica, we share the delight that the wife of a former British governor made possible. She took advantage of her position by introducing the sweet-sounding tree frog of Martinique to the woodland around Government House. Descendants of her imported amphibians still serenade there and in many other parts of the island to which they have spread. We think of the tree frogs as outdoor pets,

Ghost crabs scavenge the sandy beaches from Long Island to Rio de Janeiro, often dragging pieces of fish or Portuguese men-of-war from the drift debris into their deep burrows. Active chiefly at night, these crustaceans adapt quickly to unusual living conditions. Ghost crabs run rapidly and skillfully avoid obstacles, such as this shell of a channeled whelk. (Photo by Rex G. Schmidt, U.S. Fish and Wildlife Service)

cast off to make a living as best they can in an unfamiliar land. Ecological consequences of the introduction may be less obvious.

Cuban refugees in parts of Florida near Miami brought with them a few Cuban tree frogs, which are comparatively large amphibians of this type. Their modest calls are now familiar southward along the Florida Keys. Some individuals are tenors, and others almost falsetto, with a range of pitch through more than an octave. But the sound or sight of these nocturnal callers may give more lasting pleasure than the ecological impact of the newcomers on the community in which they now hold a place: the Cuban import seems to have reduced the area's abundance of both the green tree frog and the squirrel tree frog. It may have caused the present scarcity of the Florida cricket frog and the little grass frog, which is the tiniest of North America's frogs. Certainly no smaller amphibians are safe in the same terrarium with a Cuban tree frog, for it will eat them one after another.

We also credit an unknown Cuban expatriate with introducing a vigorous competitor of the slender native lizard that formerly ranged from Key West to Virginia, through Oklahoma to eastern Texas. So readily does this indigenous animal change color from leaf green to mottled shades, and finally to brown, that it became famous as the "American chameleon." It is a delightful neighbor with an appetite for unappreciated insects, even if its control over hue and pattern falls far short of that shown by true chameleons in the Old World. These native lizards will stand still, perform a few pushups to draw attention to the site, and then swing into view a bright pink dewlap or throat fan, stretched by a slender rod of cartilage.

The only American chameleons we could find recently during several months in Florida were in the remote Corkscrew Swamp Sanctuary of the National Audubon Society and the area along the Gulf coast north from Clearwater. Elsewhere, around Tampa Bay and Saint Petersburg and along the Atlantic coast from Pompano Beach to the Keys, the Cuban relative (*Anolis sagrei*) had taken the place of the native lizard (*A. carolinensis*). Both reach full size at a length around seven inches. The Cuban kind changes only slightly from its dark brown color, and its dewlap is orange, often flecked with brown. No one is sure yet whether cold weather north and west will leave a sanctuary for the indigenous lizard while the Cuban outdoor pet consolidates its hold on the warmest parts of Florida.

Two other lizards new to Florida appeared in sanctuary areas during the spring of 1974. Probably both were pets freed by people who no longer cherished them. Surely a wildlife sanctuary would be an ideal place to release them. Rules forbid anyone taking an animal or plant away from the park but make no mention of adding to the life of the place! One of the newcomers startled us when it climbed nimbly up the smooth trunk of a royal palm in the Fairchild Tropical Garden a few miles south of Miami. The total length may be eighteen inches—almost half of it tail. From our color films we identified the creature as the knight lizard (*A. equestris*) of Cuba. Later, when we mentioned the reptile to the director of the garden, he admitted that a small population of these Cuban imports is now well established. Neither he nor we could imagine what small creeping animals of southern Florida would nourish so big (although agile) a predator.

Our other find, a tree iguana at least four feet long, aroused our curiosity far less, for it is a vegetarian with catholic tastes. We found it wandering freely near the main visitor center of Everglades National Park. This reptile, which will defend itself by lashing expertly with its tail, attains a length of seven feet in Central America and northern South America. There it feeds on foliage and fruits high above the ground in the tropical rain forests. Almost certainly the individual in the national park had been left there by a visitor. That the tree iguana had some experience as a pet might explain why it posed so readily for photographers, some of whom approached within a few yards.

Elsewhere, such as in the British Channel Islands, we encountered at close range the descendants of some freed cage birds with a fascinating history. Although they originated in Asia, they are known as Balkan doves or collared turtle doves because they came to the Balkan portion of the sprawling Ottoman Empire as confined pets of Turkish army men. A few escaped accidentally, but a major release coincided with the wars of 1912–13, which forced the Turkish army to pull out precipitously from areas that became Albania, southern Serbia and Bulgaria, and northwestern Greece. Yet for years there, the collared turtle doves seemed ill-fitted for even city life. Despite their larger size, they suffered more than the widespread European turtle

doves from exposure in winter. The Asiatic birds pene-
trated as far north as Budapest, but those captured in that
city all had deformed, frost-bitten feet.

About 1930, the former pets began to spread northward.
No one is sure what environmental change allowed the explosive increase in range. Not until the 1950s did the use
of pesticides reduce the competition between collared turtle
doves and similar native birds, whose diet makes them
more susceptible to poison. Not until the 1960s did the
climate of northwestern Europe show a warming trend.
Perhaps the expanding cities and industries offered progressively more heat in winter. By 1955 the Asiatic doves
settled in the Channel Islands and the British mainland.
During the 1960s they began breeding and spending the
winter in Scandinavian cities and in Poland. Bird-watchers
in Stockholm and far up the Norwegian coast see the newcomers replacing local species, such as nuthatches, sparrows, tits, and blackbirds.

The collared doves do not migrate from their new locations, which now include Ireland and Spain. This is true
also at the eastern end of their Asiatic range in Japan. Instead, they congregate to share the grain put out for urban
pigeons, domestic fowl, and other birds. Officials at the
Prague zoo complain that they must increase the food
supply about 20 percent for their display birds because the
foreign doves swoop down so regularly. Sometimes the Asiatic doves build a crude nest in some sheltered corner and
lay a pair of eggs in it as early as Christmas. Two weeks
later their young hatch. Within two months the parents are
ready to start the next family. They may continue like this
into late September.

The collared turtle doves in the Channel Islands hold
daily meetings in a pine tree that overhangs a boat dock.
The males puff out their pale feathers and swivel before
prospective mates, cooing in a rhythm that distinguishes
them from other members of the pigeon family. They keep
alert for people arriving before boat time, who nibble at

snacks and toss a share to any bird that will come close. Immature doves, which lack the narrow black neck marking of the adult, take more chances with children. Perhaps the adult birds are less hungry because they have completed their growth. Or are they just too busy keeping other kinds of birds from the free food? Aggressively the doves displace the somewhat larger pigeons, with a behavior that surprises people wherever the Asiatic birds settle. None of these actions, or evidences of adaptability, would have been evident to army men who kept doves in cages. European naturalists, who know the history of these birds, sometimes suggest that the belligerent actions they show developed in captivity under the tutelage of human warriors!

Predicting what a caged bird will do when set free is almost impossible. The effect of confinement remains equally hard to imagine until it is experienced. The naturalist in charge of the Audubon Center in New Hampshire's Bear Brook State Park showed us how she emphasizes this point to young visitors. No longer has she any mammals or birds in cages for youngsters to see at close range. Instead, the empty cages have unlocked doors and a sign. It invites children to crawl inside and imagine how the animals used to feel, looking out but unable to escape.

We ourselves have twice been surprised when we carelessly allowed a member of the parrot family to escape. One was a red-crowned parrot from the arid highlands of northern Mexico. The tame bird had been given to us years after being imported legally to the United States and kept as a house pet. Its former owner wearied of the loud wake-up call the vigorous bird repeated early every morning— and sometimes at other hours as well. We saw no need to keep this parrot in a cage, for it would perch and posture by the hour on a wooden bar or a shoulder. The bird flew only laboriously and for a few yards, always losing altitude, because we followed a routine established while the parrot was young. After every molt, we scissored off the end two inches from its flight feathers on one wing. Then came the day when the parrot perched quietly beside us in the sun near the rocky shore of the Monterey Peninsula in California. The sudden appearance of an unleashed dog frightened the bird. It took off in panic, only to be caught by a gust of wind and carried, flapping and screaming, toward rough rocks over which scarcely pacific waves were crashing. We rescued the parrot, but not without wondering how different the outcome might have been had the breeze carried the bird in the opposite direction, to a grove of trees. That bird could climb magnificently in a three-handed manner with sturdy beak and clasping feet. Would it have returned to us when hungry?

Perhaps our second experience provides an answer, although the bird was of a different kind. It was one of a pair from Southwest Africa, a peach-faced lovebird. Both individuals were getting on in years, but this particular lovebird never failed to engage in vigorous flying exercises daily while holding tight to a crossbar in the cage. We took the caged pets with us to Cape Cod rather than arrange for a bird-sitter. We left the cage untended for a moment while our hostess showed us her garden. In a sheltered, sunny corner, the lovebirds seemed quite safe in their secured cage, yet in that moment a neighbor's cat shoved the cage door open, and reached a paw inside. The frailer bird merely cowered—and was safe. The vigorous one darted to freedom and perched in a tree. We tried in vain to induce our pet to return, using all the ways with which we had had complete success inside our home. For two days the lovebird stayed within calling distance of its caged mate—and call back and forth they did! Then the fugitive pet was gone, perhaps with other fliers to some person's feeding tray. In freedom it may even have survived the winter. Its cagemate, meanwhile, languished and died within two weeks.

South of Miami, Florida, and in southern California, people are getting accustomed to pairs of raucous parrots and small flocks of parakeets flying overhead. At first, we suspected that promoters of the Parrot Jungle, a first-class entertainment center in Kendall, were deliberately allowing some of their birds to fly free. Near their showplace we often noticed both the scarlet macaw of lowland forests and the green "military" macaw of mountain regions flapping overhead in a fine facsimile of their natural behavior in Mexico and southward. *Amazona* parrots, barely more than a third as big, are recognizable by their heavy bodies and rapid wingbeats; the several kinds from Latin America are distinguished by their color patterns (which are almost invisible in silhouette) and their raucous cries. Parakeets dart about so fast that they are still harder to identify unless they chance to reflect the sunlight or vocalize in shrill patterns that are instantly meaningful to anyone who had kept the particular species as pets.

Then we discovered that many of these birds no longer return to food and shelter that people provide. Instead, they find fruits and seeds on trees and shrubs that will grow in almost any warm country. One flock of canary-winged parakeets arrives regularly each morning in a tall mango tree about midway between Kendall and the Miami Museum of Science and Natural History. Excitedly the birds chatter and compete for unpicked fruits that have ripened fully. Many mangoes drop half-eaten, broken from the branches by all the commotion. When the supply of mangoes ends the parakeets abruptly shift their morning attention to a young baobab tree, in which a totally different type of fruit is ripening. Somehow they manage to cut through the tough outer covering and reach the seed-filled pulp inside. A parakeet could almost disappear through the rough side hole in cleaning out the contents.

As we watched and photographed from ground level, we thought of the geographical mélange: parakeets from northern South America, freed or escaped from a life as caged pets in southern Florida, feasting first on products of a tree introduced from the foothills of the Himalayas and then on a different tree from Africa beyond the great deserts. The baobab, in a sense, was a pet too. It was being protected and fed with special nutrients for the delight of an owner who could not often visit the part of the world from which the exotic living thing had come.

Someone, it seems, should have had the knowledge, wisdom, and authority to exclude certain birds from any importation into the United States. High on the list of undesirables would then have been an eleven-inch parakeet that is a pest in its native South America. Yet, beginning in 1968, twelve thousand of these green-bodied parakeets with gray breasts and hoods were imported legally and sold as "monk" or "Quaker" parakeets. Their homeland extends from southern Brazil to central Argentina. There great flocks of the birds travel for long distances to reach grain fields, or they settle in fruit orchards and cause disaster. Argentinians have mounted massive campaigns against parakeets that could not be driven off. Millions were shot, snared, or poisoned. Thousands of bounties were paid for pairs of monk parakeet feet. Nor was experience in foreign lands much different. Birds of this kind got loose some years ago in Yorkshire, England, and became an expensive nuisance, despite all efforts to eliminate them.

At least one crateful of monk parakeets was dropped at Kennedy International Airport in 1969. It broke open, freeing a dozen pairs from ever being caged as pets. No one expected them or any descendants to survive a winter. But they did. So did many others that escaped or were freed. The American Museum of Natural History was besieged with telephone calls from excited New Yorkers who saw parrots on their window ledges, on television aerials, at feeders, or just flying about in flocks. The following spring the news was of parrots building bulky nests of twigs—something the monk parakeet is almost unique in doing but does quickly and in colonies. Ornithologist John Bull responded to one call by going to suburban Valley Stream, Long Island, in time to see two nests being built in adjacent trees. Within two weeks the cooperating birds constructed tightly woven masses about four feet high and two feet across, with a dozen or more individual apartments in each, every apartment a one-room space with a separate doorway.

Now all five boroughs of New York City have monk parakeets nesting. Two broods per year seems average, with two to four eggs (and nestlings) in each brood. There is no way to be sure whether birds from this source or from other releases of pets provided the start for populations now well established in New Jersey, Maryland, Virginia, western New York, and Michigan. To let Californians see monk parakeets building their spectacular nests, officials at the San Diego Zoo freed fifteen of the birds in Balboa Park. When nearby residents complained and the California Department of Agriculture filed a protest, the zoo staff set out to recapture the birds. Apparently thirteen were caught and the remaining two disappeared. Whether or not they perished may not matter, for more monk parakeets were soon seen in Anaheim, Santa Ana, Pomona, and elsewhere in California.

No one can yet predict how widely these introduced parrot pets will spread or how they will affect the many different living communities. So far the monk parakeets have shown no tendency to usurp the nesting sites of other birds or to remain so constantly at feeding stations that other species have no chance. The few instances where the monks have harassed a pet dog or cat seem exceptions, although they may represent a behavior that will become more important as the birds increase in number. A parakeet that will eat beetles might do worse than to dine regularly on foods prepared for dogs and cats. Their diet, upon which people in the United States now spend almost $2.5 billion annually, would be well worth fighting for.

Despite the huge investment in creature comforts for pet dogs and cats, a great many of these animals leave home or are cast off to fend for themselves. The dogs revert to their sociable habit of hunting in packs, whereas the cats continue to be the "wildest of the wild animals," traveling as individuals, as though all places in the world were alike to them. Sociologists now study the territorial organization of urban canines that haunt the city streets at night, seeking edible refuse. Rural people decry losses of livestock to packs of feral dogs and claim about $5 million in damages to cattle alone each year. Sportsmen rage at the "hounds of Hades" that track down and kill deer, destroy wild turkey nests, and attack most other kinds of game and, occasionally, people. The dog warden of Wake County, North Carolina, regards as average a year in which he is called upon to dispose of thirty-five thousand dogs that are running wild and doing damage in the countryside around the city of Raleigh. Some of these animals prove to be licensed pets, on vacation from their owners. Wildlife managers at Auburn University, Alabama, found that they could distinguish between the tame and the free-ranging dogs in their area by the vicious behavior of the long-feral animals when trapped; pets that had only recently left home reacted in a docile fashion to being confined again. But when released with compact radio transmitters attached so that scientists could trace their movements hour by hour, the tame dogs and the wild ones roamed indistinguishably. They foraged in uplands during cool weather and preferred the shady floodplains in summer. Hunting chiefly at night, they covered between a third of a mile and five miles each day, usually within a home territory totaling less than four square miles (twenty-five hundred acres).

In some states, the original purpose in charging an annual license fee for pet dogs was to derive a fund from which the state could recompense sheep owners for dog damage to their flocks. If the fee seems nominal, unaffected by inflation, its stability reflects the declining number of sheep in these states and the infrequency with which a claim is made against the fund. In states where sheep are numerous, the ranchers prefer to ignore the feral dogs and pretend that wild coyotes are entirely to blame for dead lambs or adults. This enables them to obtain federal funds, for federal procedures do not require detailed proof that a particular predator is responsible. Proposals have been made for establishment of either a bounty system or a program of predator extermination, either of which would bring money to local people, over and above the payments for specific losses among sheep.

The beneficiaries of a bounty system are likely to boast about their prowess in collecting bounties and to be offended if newcomers to sheep country question the need for this particular use of public funds. Reliable information on sheep kills and the coyotes' role are seldom sought where prejudice, profit, and emotion influence human attitudes toward these wild animals. An out-of-state student from New Jersey, John Litvaitis at Northland College in Wisconsin, reacted strongly (but, wisely, with few comments) when he saw a newspaper story accompanied by a picture of a huge pile of dead coyotes, about the coyote bounty. "What a waste of an animal," he thought to himself. "There has to be a better way!" But after making a study of Wisconsin's bounty system as a project in a wildlife management course at Northland, Litvaitis transferred to the University of New Hampshire to pursue graduate research on coyotes in New England. He concludes that "they almost seem to benefit by experience, since they continue to produce new generations of smarter, more adaptable coyotes" wherever mankind persecutes this predator.

The state legislature in Missouri decided to end the bountying of coyotes a few years ago and to urge sheep ranchers and farmers instead to try setting traps for predators within the sheep enclosures. Any coyote that got caught could be shot. But the sheep farmers found the traps more a nuisance than a benefit, for every morning without fail the traps had to be checked. A neighbor's dog might be inside. Rarely was a coyote bothering the sheep. Finding and dispatching any actual predator proved far more effective than blanket control had ever been.

In Montana, where dogs are relatively few and coyotes numerous, 1974 saw two studies to investigate the pros and cons of predator control, which is a euphemism for extermination. The Montana Department of Livestock conducted one study in twenty-one counties where depredations had been reported. In six months, a poisoning campaign killed 136 coyotes, 70 foxes, 9 skunks, 2 raccoons, 1 badger, and an undisclosed number of feral dogs. The total cost approximated $40,000, "not including salaries and travel of employees who trained applicators in the use" of the poison. This amounted to $294.11 per coyote taken, or $194.12 per coyote or fox. The cost per dog was best left uncalculated, for fear of raising a storm of protest.

The U.S. Fish and Wildlife Service paid for the other investigation, which was initiated by the University of Montana. The aim was to record all sheep losses and identify the cause of each one, if possible, on an 8,500-acre ranch in the Bitterroot Valley south of Missoula. On this ranch the previous year, 61 coyotes had been killed without destroying those individuals that were responsible for fatal attacks on sheep. During the six-month study, no more coyotes were taken. But the same survivors, plus perhaps a few immigrants, were responsible for killing 429 sheep, of which 364 were lambs. Feral dogs killed 2 sheep and wounded 11 more; 2 lambs were taken by golden eagles; and 66 other sheep of various ages died from a variety of causes, some of which could not be identified. The researchers concluded what everyone knew: that "coyotes under certain circumstances do kill sheep." Their study did not reveal, however, whether more than 2 coyotes had become sheep-killers, whereas 2 of the scarce dogs had been caught and killed in the act.

It is easy to see where so many wandering dogs originate and why bird-watchers find feral cats responsible for more wrecked nests and dead nestlings than any other kind of predator. Investigators for *Time* magazine summarized the statistics they gathered for the issue of Christmas 1974 as showing in the United States 100 million dogs and cats, which reproduce at the rate of 3,000 per hour, compared to about 215 million people adding 415 human babies per hour. While providing the basis for a pet explosion, these dogs and cats daily add to the nation's waste an estimated 4 million tons of feces and 42 million quarts of urine, chiefly on city streets and parks—a public embarrassment of enormous size.

How changed the ecological relationship between mankind and household animals is now, compared to when the dog became the "first friend" of our species and the cat moved in where it could enjoy a place by the fire and have warm milk to drink! No longer does the pattern follow the form that originated long ago in both the Western and Eastern centers of civilization.

Modern anthropologists, who realize how commonly African lions usurp in early morning the prey that hyenas and wild dogs have killed the night before, credit Ice Age huntsmen with being the lions of their day. They watched until a pack of wild canines pursued and downed some large beast and then drove away the killer to claim the meat. After removing the edible muscles, the liver, and perhaps the brain and the bone marrow, the huntsmen abandoned the remains for the dogs and any scavengers that might come along. To this degree, people depended upon dogs and appreciated their ability to run down and kill game.

Later in Europe and the Near East, dense forests grew up as the glaciers receded. The trees impeded travel and reduced the food available to both mankind and wild dogs. People settled in small groups where they could maintain a clearing and catch fish all year to supplement whatever the best huntsmen could bring home. A human community became a source of wastes that hungry dogs could eat. It provided also a testing ground for primitive agriculture and a center in which the culture identifiable as Mesolithic (Middle Stone Age) could evolve. A hunter returning almost empty-handed to his family might soften their disappointment by bringing a pup or two taken from the den of a wild dog. Did the pup become a pet, an early-warning alarmist contributing to the safety of the human community? Did the tamed animal go along silently with the huntsmen and sometimes help in finding game or holding it at bay? It seems likely that the breeding of dogs in captivity developed at this stage. Soon the Greeks could imagine a formidable Cerberus guarding the gates of Hell and the Britons could begin selecting the ancestral line that, by Julius Caesar's time, had become mastiffs—canine mercenaries to bring back to Europe as dogs of war.

In the Far East, the climate seems always to have been excessively dry. First, it provided little moisture from which the Ice Age could produce great glaciers. Then it supported no deciduous forests to crowd people together. Instead, culture arose along the shores of rivers amid grasslands, where irrigated crops of cereals could support a growing human population. Rather abruptly, as the weather warmed after the Pleistocene, the descendants of hunter-gatherers became settled farmers. They went through no transition period, no Middle Stone Age. And their attitude toward dogs fragmented in several directions: toward regarding wild ones as undesirable predators with the one redeeming value that they could be killed and eaten; toward selecting compact individuals as a breeding stock from which to develop miniatures such as the Pekinese; and toward breeding larger animals that would guard the homes of the affluent few, as a chow dog will. This tradition continues in Hong Kong and Taiwan, possibly also on the mainland of China and in the Chinatowns of large American cities, where dogs are scarce and sidewalks tend to be clean. There stray dogs and cast-off pets have no place.

The mouse-catching abilities of the cat have been appreciated ever since mankind began to store dry grains as a food bank. Unlike most predators, a cat will lie in wait for

hours to catch a mouse, play with it as a progressively injured toy, perhaps bring it home as a limp trophy, and then
turn to a preferred meal in a favorite dish. Several of the
world's extinct birds are known only from specimens that a
house cat brought in. We ourselves are frequently amazed
by the array of inconspicuous animals our neighbors' cats
catch, only to have the bodies confiscated and taken to the
nearest naturalists for identification. If not for the cats, we
might never suspect these creatures live nearby.

The Egyptians, of course, revered the cat as a sacred
beast, one able with its shining eyes to see in the dark and
into the future. One pharaoh boasted that he had twelve
thousand servants merely to attend to the needs of his
sacred cats. Cats were elaborately mummified, or their
skins stuffed and studded with shining jewels to be interred
with the body of a major personage. Probably all of these
Egyptian cats were derived from the African wildcat (*Felis
libyca*), whereas the Siamese breed from the Orient may be
descended from the steppe cat (*F. manul*). Yet the natural
reluctance of cats to follow human guidance has hampered
all efforts to breed special strains. Cat fanciers can neither
offer a neat catalogue of distinct cat kinds to correspond to
the 120 recognized types of dogs in the files of kennel associations nor conquer any cat's determination to go off
alone, to choose its prey, to practice its inherited reflexes.
Cats cast themselves off, even if their owners do not.

Today more than five thousand different kinds of wild
animals are caught, kept as pets, and often discarded far
from their place of origin. That represents fully 10 percent
of the known species of vertebrates, plus a generous assortment of crabs, crickets and other insects, scorpions and ta-
rantulas (among the spiders), octopuses, and many types of
snails. The logistics of transporting creatures with so many
different requirements of food, companionship, and space,
let alone breeding them in captivity, impose a formidable
challenge to human ingenuity. So does preventing their escape, which can entail far more loss than the investment in

procuring the pet itself. The cost to human society may easily exceed that from having a child play with kitchen matches of the "barn-burner" type. The importance of prompt action, equivalent to that of an efficient fire department, cannot be overlooked.

We have no reason to regard as unique or peculiar the Miami child who returned from a holiday in Hawaii recently with a giant snail, alive, concealed among dirty socks in his luggage. The airline officials had no idea what their jetliner was carrying. The boy's mother foresaw no further excitement after her son complied with her command, "Get that thing out of the house!" How would she recognize that this kind of snail is out of control in Hawaii, having been introduced there by a Hawaiian woman of Japanese extraction? The Hawaiian woman had taken a fancy to the snail while on a visit to some islands in the South Pacific that the Japanese military forces had held during World War II. How could she guess that continuing devastation to plant crops by snails on those islands now exceeds all the damage done by military action during the war? She may not even have known that the Japanese introduced the ancestors of the snail, which they encountered first in Burma. The Burmese encouraged the Japanese to take some of the snails to Japan, where the soft bodies might be minced and tried as an aphrodisiac.

Only later did a sleuthing biologist discover the origin of the land snails in Burma. Their ancestors were released in the Chouringhie Gardens just outside Calcutta by W. H. Benson, a British conchologist who had tired caring for them after admiring and collecting them on the island of Mauritius in the Indian Ocean. The snails on Mauritius were in the garden of a friend and fellow shell-collector, Sir David Barclay. Sir David had brought them from East Africa, where the giant snails are native. From Natal to Somalia and on the big island of Madagascar, the people appreciate the land snail *Achatina fulica*. They gather those that are approaching full size, with an attractive brown shell as much as five inches long and a live weight of almost a pound. East Africans cook and eat the snail bodies and carve the shells into utensils of many kinds. And if the people abandon these cultural traditions in favor of those from other lands, the giant snails still are kept in check by the civet cats, a land crab, one or two carnivorous snails, and a few kinds of beetles that prey on the young snails before the shell gets too thick or the creature is able to flood its soft surface too copiously with mucus. Elsewhere in the world, no such array of neighbors holds down the rate at which *Achatina* can transform vegetation into more snails.

The giant land snails that began their travels in suitcases of visitors almost 130 years ago have been making unpredictable landfalls ever since. Usually they become well established and out of control before the consequences of their arrival are appreciated. So far every arrival in America and Europe has been destroyed as quickly as the ones in North Miami during 1969 and again in 1974. Each time a child loses an intriguing pet, but the horticulturalists and the shippers of fruit and vegetables avert a certain calamity. Just one gravid snail is all that is required to start each *Achatina* infestation.

When we stop to think about it, we realize that an animal is expected to adjust quickly to an unfamiliar diet when it becomes a pet in a strange land. A parrot is offered sunflower seeds in spite of the fact that this food is of North American origin and the parrot almost certainly from some other continent. A dog or an owl is fed defrosted horse meat, although this nourishment would ordinarily be unavailable to it. And any aquarium fish that prefers food of animal origin is expected to be satisfied with freshly hatched young from brine shrimp eggs—a purchasable product that can be stored dry on a shelf—or with mealworms, which no fish would be likely to encounter in nature. The new habitat that the pet is offered reflects mostly what is readily available. Who would worry about providing privacy for a goldfish?

Yet when the pet dies or must be discarded, what happens to the artificial environment in which the creature has been living? The pen for the guinea pigs may simply be dismantled and burned in the fireplace. The cage for the bird might at least be cleaned and put in storage. Should the green plants in the fish tank be dumped on the compost heap, flushed down the sewer, or emptied into the nearest pond? If the fish are to be released, why not leave them in their familiar surroundings? And *that* decision is so common that aquarium plants have been distributed as widely as pet fishes.

An unknown fancier of aquarium life probably added the common North American waterweed *Elodea* to the British scene. Its presence was discovered in 1842 by Dr. G. Johnston of Berwick-on-Tweed in a small lake on the grounds of Duns Castle in Berwickshire, Scotland, close to the English border. Soon the bright green foreigner turned up in canals, ditches, rivers, and ponds all over the United Kingdom. Scientists at Cambridge University made the mistake of introducing some as a curiosity into the River Cam. Quickly the weed clogged the channel, interfering with rowing crews from the university and requiring the services of extra horses to haul barges for commerce. Laborers cleaned out the plants but missed a few fragments. The waterweed regenerated entirely vegetatively, since only staminate plants reached the Old World. British aquarists began adding short lengths of the weed to their fish tanks. There it is decorative, a fine generator of oxygen, and easy to trim before it occupies too much space.

Waterweed is still a pest in many European streams. It has been discovered in Tasmania, New Zealand, South Africa, and other distant waters. Cattle and some waterfowl learned to eat it, but mankind can claim no credit for getting the plant under control. The crest of its abundance passed in the 1860s. Yet experiences with *Elodea* led to no urgent recommendations regarding aquarium plants, let alone any worldwide legislation.

Floating weeds that are favored in the tropical-fish and garden-pool trades now cost millions of dollars in control programs that offer no real hope of extermination. Most famous are alligatorweed and water hyacinth, both from eastern South America. There various insects and sea cows (manatees) keep them suppressed. South American beetles now control alligatorweed in some southern states, but water hyacinth continues troublesome. Each water hyacinth forms a rosette of bright green leaves. Their shiny blades, four to six inches wide, are buoyed up by gas chambers in the bulbous petioles. Fine black roots trail into the water. They grow also from horizontal stolons that produce new small plants. In season, mature plants send up stalks bearing a dozen pale magenta flowers, each marked handsomely with yellow and dark purple. Self-fertilization sets seeds if no insect comes with pollen; ripe seeds are released under water. They may lie dormant for twenty years before germinating to produce new floating plants.

Blossoming water hyacinths reached New Orleans in 1884 from Venezuela, imported by members of a Japanese delegation to an international cotton exposition, to be distributed as souvenirs. By 1890 the weed had become naturalized along the lower Mississippi and in Florida and had begun to spread northward along the Atlantic coast. It can be blown along by wind and tolerate seawater for a week or more. By the turn of the century, it was as far north as Virginia and as far west as California. In many places it interfered with drainage and with navigation.

Floating masses of water hyacinths plague the East Indies. They originated from a misjudgment in 1894 at the famous Buitenzorg (Bogor) Botanic Gardens in Java. A display of water hyacinth, imported for the purpose, grew too luxuriant and was dumped into the nearest river. Infestations developed at several sites in continental Southeast Asia by 1902 and in Ceylon by 1905. By 1956 it reached the Nile in the Sudan. Now the seductive plant is a major pest in the Congo basin of Africa, Celebes, Borneo, the Philippines, Guam, and northeastern Australia; it is well known in China and Japan. Everywhere it shades native vegetation to death.

Floridians and the U.S. government spend millions of dollars annually to suppress water hyacinth. Spraying the floating mat with weed-killers destroys a majority of plants, but the remainder turn an unsightly brown (as they do with frost) without losing the growth centers from which the infestation renews itself. Conservationists would like sea cows to do the work gratis. But these hulking animals cannot tolerate either the chill of winter in fresh water or the sharp propellers of motor boats that race along any open channel. Dams intended to limit the inland flow of salt water during dry seasons block the normal migration of sea cows between refuges seaward in winter and warm-weather feeding grounds among the hyacinths.

This experience has had little effect on the managers and patrons of the ornamental aquarium trade. Water hyacinths are readily available in almost every state. We repeatedly find a small colony of them in New Hampshire where they are discarded in an abandoned reservoir. Winter kills the discarded hyacinths, but not those in someone's heated aquarium.

Economists and engineers did not calculate upon the introduction of floating weeds when they planned the great Kariba Dam across the Zambezi River, where it forms a boundary between Rhodesia and Zambia. The planners weighed chiefly the costs of construction against the benefits to be gained from electric power. The new lake would extend upstream almost to Victoria Falls and require the relocation of people from areas to be flooded. "Operation Noah" would salvage some of the displaced wildlife. Yet no more than one aquarium full of pet fish and attractive water plants was enough to upset the scheme. Was it a humane contribution by one of the foreign consultants whose residence at the dam site ended when the dam gates were closed? No one knows—or will tell. Or did the transformation of a great river that rushed along and scoured its banks into a tranquil lake provide the habitat for plants already there?

The weed that first colonized the Kariba impoundment was not water hyacinth but a much smaller plant: the water fern of northern South America, Central America, and the Antilles. It appeared within months after the Zambezi began to rise behind the Kariba Dam, an event celebrated in December 1958. By 1962, colonies of water fern extended along most of the shoreline, and vast green mats of it covered as much as a third of the entire lake surface. Winds propelled the mats from place to place and aided in their consolidation. So did recruitment of native water purslane and a Cuban bulrush, for these provide a weft of interlacing roots and horizontal stems that strengthen the mat until a man can walk upon it.

Lake Kariba quickly acquired another aquarium plant—water lettuce. It floats in great masses on its own or joins up with water fern. Each year, part of the profit from the sale of electric power from generators at the costly dam must be spent in raking up and destroying this floating vegetation. It multiplies ceaselessly on the surface, with no annual frosts to set back the continued reproduction. No sea cows are available to give assistance free. The cost of limiting the escaped pets to a tenth of the lake area seems likely to continue as long as the dam stands and the river cannot wash them out to sea.

Sporadic news of aquatic weeds includes some good, some bad. Yet the accounts rarely point to any connection with the wet-pet trade. Water hyacinths now thrive in several polluted rivers of Europe and America and show an unexpected ability to clear them of some toxic substances, including the salts of heavy metals. But these plants are spreading too in Lake Victoria and Lake Tanganyika, which are the largest in Africa. There they shade the shallow water and reduce the production of fish—a source of protein needed by people who live nearby. Water hyacinths among the floating masses of papyrus reeds severely retard the flow of the White Nile through the 2,500-square-mile marshland known as the Sudd in the southern Sudan. To combat this development, the Sudanese and Egyptian governments plan to invest $350 million in a 180-mile canal. Its proponents expect the new waterway to be completed by 1980, to be relatively easy to clear of weeds, to double (at least) the flow of the Nile to the Aswan Dam in Egypt, and to allow diversion of irrigation water into more than 3,900 square miles of potential farmland to be reclaimed in the Sudan. Opponents discount the claims that breeding places for disease-carrying mosquitoes will be eliminated. Scientists can predict the prompt spread of severe infestations of blood flukes in all irrigated areas, followed by a disastrous change in the whole pattern of rainfall in this part of equatorial Africa as the Sudd is drained. Neither the hydraulic engineers of the Sudan and Egypt nor the primitive Nilotic people who have so long maintained their self-imposed isolation along the myriad waterways of the marshy Sudd can foresee the endless consequences that are sure to follow any rerouting of the great river.

So far, the damage caused by cast-off aquarium plants and pets outweighs the occasional benefits. Enough native species are crowded out by each introduction to make the situation ominous. Perhaps the natural residents need a different form of aid. It could be as simple as the discovery of a gainful method of utilizing the vigorous foreign pests.

Already some progress has been made. In Florida, where studies at the state university failed to detect any dangerous chemical substances in water hyacinths, engineers are testing a waterside machine that can press out the nineteen hundred pounds of moisture from each ton of the floating plants. It readies the hundred pounds of residue for sun-drying and conversion into "hay" or pelletized feed for livestock. The product offers about the same proportion of proteins, lipids, mineral nutrients, and fiber as the grasses ordinarily grazed by sheep and cattle. Certainly the plants recover quickly from harvesting and provide a self-renewing resource. At Bay Saint Louis, Mississippi, the National Aeronautics and Space Administration is readying a large-scale sewage filtration facility using water hyacinths to treat wastes from about ten thousand people. As the contaminated plants grow, they will be harvested and fermented in tanks to yield methane gas for fuel and a sludge that can be used for fertilizer. No one foresees any need to encourage the waterweeds, for they thrive without chemical supplements, irrigation, or cultivation techniques. Supervision to keep up with their prolific growth may succeed in converting them from pests to crops.

4

Something New to Eat

AT THE GALA DINNER of the prestigious Explorers
Club, held annually in the posh Waldorf-Astoria Hotel in
New York City, world travelers in tuxedos circulate around
the table of hors d'oeuvres. There they chat and smile
while reading the labels and sampling an almost unpredictable assortment of delicacies from everywhere: braised
filets of python, thin squares of zebra brisket, small slices of
steamed rump meat of the goatlike Himalayan tahr,
kangaroo tail, pickled shark fin, cocktail crackers smeared
with caviar from the white sturgeon of the Black Sea, tiny
sausages of armadillo meat, Aleutian salmon liver with tundra berries, fried Katanga termites, spruce-bark bread,
stuffed Liberian breadfruit, mashed cactus pears, and
cubes of meat from an extinct mastodon that was fresh-
frozen more than ten thousand years ago in the arctic tundra but defrosted recently. Some of these oddities prove to
be amazingly flavorsome.

Part of the pleasure in travel comes from trying unfamiliar foods, prepared by people who know how to bring out
the best in the local product. It might be an ostrich egg
scrambled—equivalent to two dozen hen eggs—somewhere
south of the Sahara; or a penguin egg, hard-boiled in

Tierra del Fuego; or a strip of sun-dried flesh ("biltong") from zebra or impala, bought neatly packaged from a rack in a Johannesburg market; or a freshly poached side of oil-fish from waters four thousand feet deep off the coast of Madeira, where the natives call the sharp-toothed creature an *espado* or an *escolar,* according to minute differences we could not appreciate. The memorable morsel might be Mississippi catfish with hushpuppies in the Deep South, french-fried strips of calamary (squid) on Catalina Island off the California coast, a delectable "steak" of abalone imported from the Mexican coast, or "breast" of freshly dug geoduck clam sauteed in butter, as a special treat from the mudflats around Puget Sound. We tend to savor the memories of unusual foods from far and near, whether T-bone steak of caribou in the Canadian northwest or a serving of prime ribs of American bison somewhere in Montana. We think of them in the same category as items that strained our palates far more, such as ancient eggs in Singapore and camelburgers in Cairo.

Yet there is another side to the same experience: the often desperate hunt for a familiar food far from home, where the local meals progressively evoke a kind of home-sickness. We recall with special affection the relief we found from a steady diet of rice and beans in Latin America, fine though they were. We needed an occasional dish that offered a flavor and texture we would choose in New England. In Panama, this remedy took the form of a can of Danish ham and a frozen package of asparagus spears. Near Tegucigalpa, Honduras, the magic lay in an occasional liter package of pasteurized pistachio ice cream. In Egypt, it was the plain hamburgers in the grill room of the Nile Hilton Hotel. We suspect that in Hong Kong or Tokyo we would soon find ourselves looking for a Mc-Donald's!

We feel confident that our interest in the unfamiliar and our longing for our customary diet are common to people everywhere. Their satisfaction is at least partly responsible for the arrangement made by seafood restaurants on famed Fisherman's Wharf in San Francisco. They have live Atlantic lobsters shipped to them almost daily by air from Maine. This is ten times as far as the nearest source of spiny lobsters, which are fished for commercially from the Monterey Peninsula of California to the Gulf of Tehuantepec in Mexico.

Recently, the continued importation of lobsters from the Atlantic coast has been identified as the source of mysterious additions to the fauna and flora of San Francisco Bay. The restaurateurs certainly never let a live lobster get away after having it airlifted from Maine. But each shipment of lobsters comes packed in wet seaweed, to keep the marine animals from drying out and dying on the way. The lobsterman can easily pick up enough handfuls of seaweed, such as bladder wrack and rockweed, to safeguard his catch. This practice costs him only a few minutes' time and no cash outlay. It avoids any need to have the lobsters carried in tanks of seawater, with an aerating system, which would be heavy and bulky as well as susceptible to breakdown, all adding to the expense of shipment. Only one detail is overlooked: the fate of the wet seaweed when it reaches California and the lobsters are removed. The restaurant staff simply tosses the wet seaweed over the edge of the wharf, to be carried away by the currents and tides.

Both bladder wrack and rockweed grow with one end attached to the rough surface of some rock, while the branching bladelike parts of each plant either lie limply when exposed to air by a receding tide or fan out in tidewater, buoyed up by their gas bladders. New England and northern European beachcombers are used to finding doomed seaweeds of both kinds. Seaweeds are torn from the rocks by storm waves and tossed on the shore. Those that a lobsterman uses for packing material float easily in San Francisco Bay and drift about in all directions. Along with the seaweeds go any other kinds of life that cling to bladder wrack and rockweeds. These survivors get dispersed widely, and some of them find a new habitat that fits their needs.

The first of the East Coast animals to attract attention in San Francisco Bay probably came hidden among similar packing materials used for a shipment of shellfish from farther south, at a time when no one had any idea that the transcontinental leap could be made. But starting in 1958, naturalists began finding channeled whelks, which normally occur only south of Cape Cod, Massachusetts, along the Atlantic coast to northern Florida. These snails in San Francisco Bay may attain a length of more than seven inches, just as in their native waters, and produce long strings of egg-packets that are often called sea necklaces. Each of the several dozen packets protects as many as a hundred eggs, from which hatch snails more than an eighth of an inch long.

Western conchologists who reported the immigrant whelks generally welcomed the addition. Although no one might choose to make conch chowder or a nourishing soup from the creatures, the whelks at least would provide convenient dissection material for elementary classes in zoology. Not even scientists who knew the eating habits of whelks inquired what they might feed on. Along Atlantic shores, each big channeled whelk kills and devours one big bivalve each month. Smaller individuals, which are more numerous, take a corresponding toll of smaller clams. An abundance of whelks in San Francisco Bay had to mean fewer bivalves of fair size.

At least one bivalve that is appreciated on the East Coast came west at about the same time. California shellfishermen began picking up an occasional hard-shell clam, which New Englanders call a quahog even if they market it as a littleneck or a cherrystone. Once its shells were the source of wampum, the perforated discs that served Indians as decorations on necklaces and as small change. Was the hard-shell clam an unauthorized introduction made deliberately? Or did some young ones arrive accidentally, as contaminants in an unrelated enterprise?

The existence of an undiscovered transport system from east to west across America became more likely in 1966, when shell collectors began to find the common European periwinkle snail in shallows of San Francisco Bay. This is the favorite "winkle" of England and of western European coasts, where people harvest the small snails by the bushel and sell them, freshly cooked, on street corners. The custom has not spread noticeably to the New World, but European periwinkles did. Many New Englanders believe that this is one token left by the Vikings who discovered "Vineland," probably along the Nova Scotia coast. Perhaps they carried a supply of periwinkles with them in their long boats. Shells of these snails are among the campfire remains that archaeologists have uncovered in eastern Canada, most with a date in the eleventh century A.D., according to measurements made with the radiocarbon technique. Since then, periwinkles of the European kind have apparently spread northward to Labrador as well as southward. In historic times, the common European periwinkle has traversed the Cape Cod canal in Massachusetts and extended its range to Maryland. Anywhere along the intervening stretch of coast, a few could easily have been concealed within a mat of seaweed—a favorite habitat—destined for San Francisco.

A biologist from Temple University in Philadelphia, Richard L. Miller, noticed some rockweed (*Ascophyllum nodosum*) being tossed into the water by an employee from Fisherman's Wharf. Miller began to wonder what kinds of life, in addition to the seaweed itself, might be freed into West Coast waters simultaneously. He examined samples from several shipments and found thirteen different kinds of animals (including periwinkles) and seven other species of seaweed. He realized that his list was far from complete. A peculiar assortment of Atlantic denizens was receiving a chance to establish themselves in an unfamiliar place.

Quite often the most deliberate transplants fail. Shellfishermen have tried for years to get a population of Atlantic lobsters started in the eastern Pacific. Equally unsuccessful has been the attempt to establish California abalones on the rocky coasts of New England. On the other hand, the Japanese giant oyster grows well along the shores of the eastern Pacific, although it must be reintroduced yearly because it does not reproduce under the new conditions. The giant oyster, which is raised commercially from northern California to British Columbia and is regarded as having considerable economic importance, has also been transplanted in French waters. After nearly thirty years' experience with it on the American West Coast, oystermen in the Canadian province contracted to collect a generous sample of live spat—the free-swimming larval stage that eventually settles to the bottom and becomes fixed for life—to go by airfreight to the west coast of France. The first of these shipments arrived in the spring of 1972 and started a new business.

More than a new food came with the Japanese oysters. Two different oriental snails accompanied the shipments to the American West Coast, where they settled down to continue their natural way of life—as oyster drills. Both types of snails have the adaptive ability to rasp a circular hole through the shell of a bivalve and then reach a rasplike radula inside to get the bivalve's flesh. On an oyster bed or a clam flat, they wander about freely, destroying large numbers of shellfish. They thereby reduce the commercial harvest and kill many mollusks that are important in the community of nature, if not to shellfishermen.

A large oriental seaweed turned up on the California coast the year after the first giant oysters were introduced. The weed grew prolifically, spreading north and south at a truly astonishing pace. In less than forty years, its range in the eastern Pacific extended from the Gulf of California to Vancouver Island, a distance of about twenty-five hundred miles. In many places it formed dense stands, many plants exceeding thirteen feet in length. It competed with the commercially valuable kelps, from which the harvesting companies derive extracts (alginates) that are used in skin lotions, ice cream, and other products, to give body and a smooth texture.

Within a year after the first introduction of giant oysters on the French coast, the oriental seaweed was discovered on the shores of the Isle of Wight. At first, the only indication was a few scattered strands in a lagoon. Then a large population turned up in Portsmouth Harbor. A task force was dispatched to gather up and destroy every trace of the potential pest. Two months yielded almost four metric tons. A patrol has since been authorized to clear away any sporelings, and it does so every two weeks. Not a giant oyster is harvested by a British shellfisherman that might be taxed to pay for all this labor and supervision. Yet the British fear that if the seaweed gets established, it will change the ecological community along the shores, interfere with boat operation, and block the intake lines of cooling-water systems that now function smoothly.

Our experience in North America is scant preparation for an understanding of the bickering, bitterness, and failure to take concerted action among people in the mother countries of Europe. Canada and the United States agree on management of fur seals and salmon. Mexico joins the U.S.-Canadian program for migratory birds. This seems so admirable and civilized that we expect neighborly consultation between all abutting nations. But elsewhere nationalism blocks comparable cooperation.

The French are now testing a coarse kelp (*Macrocystis pyrifera*) from the Southern Hemisphere as a possible addition to their coastal waters. It might provide a ready source of alginates and lower the cost in France of this food additive. Everyone admits that the kelp could easily spread to the coast of Britain, introduce problems for navigation, and completely alter ecological communities. No one can guess what other kinds of life, more likely detrimental than advantageous, would be introduced along with the exotic kelp. Arguments against the introduction are the same as those marshaled in 1950 when a similar idea appealed to members of the Scottish Seaweed Research Association and was squelched. Today the British are asking the French not to introduce the giant kelp, channeling this humble request through the International Union for the Conservation of Nature and Natural Resources (IUCN), which has head-quarters in Switzerland. The IUCN has more prestige than power. In the February 1975 issue of the IUCN *Bulletin,* the situation was summarized with a statement that records on file show "no known case where it has been proved pos-sible to eradicate an unwanted plant pest," whether in the water or on land. So strong a statement should warn in-telligent people against risking the high costs of unwise ac-tions.

That the addition to an ecological community is incon-spicuous is no guarantee that it will not develop a major im-pact. Yet the need for quick remedial action and the person who made the unwise introduction are likely to remain un-noticed. Today no one can be sure how the Asiatic clam got into the Columbia River in Washington. Americans of Asiatic ancestry do occasionally import a small shipment of these creatures by airfreight and serve them on festive oc-casions. To Orientals, they are "good luck clams," a delicacy to be eaten freshly cooked. Did someone at Richland or Wenatchee have a few live ones left after a party and drop them in the river? Shell collectors discovered the clams in 1938 and reported them to other hobbyists as a new addi-tion to the mollusks of the West. Did anyone transfer a few to other streams? No one admits it. Yet quickly the same foreign bivalve appeared in the Willamette River, the Sacramento River, the San Joaquin, and their various tribu-taries and then the Colorado River basin and all the irriga-tion channels that are connected to it. Up the Gila River, the Asiatic clam reached Arizona.

The small clams fairly studded the bottom of the cement-lined canals near Parker Dam in 1962 when the Bureau of Reclamation decided to clean them out. Six months later the clams were back, as many as 275 to the square foot where the bottom of the canal held a layer of heavy gravel, sand, or some decomposing organic matter. Construction foremen began to wonder whether the clams traveled unnoticed in truckloads of gravel or sand taken from one river and dumped near another. One man, more experimental than most of his colleagues, closed up a handful of Asiatic clams in a small jar and kept them without food or water for five days. They were still alive when he dumped them back into the river.

By 1961 the little clams managed to get east of the Continental Divide. They turned up on the Ohio River near Paducah, Kentucky, and soon spread to Cincinnati, Ohio. There in 1963, *Corbicula* was the dominant animal on the river bottom. The following year it was scarce, as though the unusually cold water of the intervening winter was too much for its temperate needs. But nothing suppressed the population of these clams in the Tennessee River, the Cumberland River, the Green River, or streams along the Gulf Coast in Louisiana. By 1963 they were in the Mobile River of Alabama and, the following year, in the streams of the Florida panhandle.

Probably the Asiatic introduction was at Charleston, West Virginia, as early as 1961, since the shells of those discovered in the summer of 1964 already showed several growth lines and had attained a length of 1⅛ inch—essentially full size. One of the nation's leading malacologists, William H. Heard, published a prediction: As soon as the Cross-Florida Barge Canal was finished, *Corbicula* would be into the Saint Johns River and other streams emptying into the Atlantic Ocean. Meanwhile, the infestation spread southwestward, into New Mexico and Texas in 1966, into the Rio Grande soon after, and then into Mexican tributaries and irrigation canals from Nuevo Laredo to Mexicali.

The Cross-Florida Barge Canal project was abandoned before completion. But the Asian clam got into the dredged channel of the Caloosahatchee River west of Lake Okeechobee in 1969, with all of the interconnected waterways of South Florida available to it. It also got into the Altamaha River system of Georgia and by 1971 had moved downstream almost to the Atlantic Ocean. By 1972 it was in the Savannah River, the Pee Dee of northern South Carolina, and the Delaware between Trenton, New Jersey, and Philadelphia. In the next year it turned up in the artificial impoundment known as Stow Lake in Golden Gate Park, San Francisco; in the effluent channel of an electric power station in northeastern Iowa; and a long list of other sites. Litter-disposal crews in central California discovered that every one of fifty discarded metal cans and bottles picked out of an irrigation canal was about four-fifths full of Asiatic clams. Some of these had grown after entering the container and were too big to pass out through the opening again. Sunken automobile bodies were filled to the windows with a mixture of mud and clams. And in one place where contractors had used sand from a river as part of the mix for concrete, so many Asiatic clams had burrowed upward to the surface before the mixture set that the work had to be done over again. It was pocked with exit holes and weakened by all the burrows. How could a small, slow-moving mollusk travel so far so quickly and reproduce at such a fantastic pace?

Conservative scientists began to wonder and inquire in print whether this exotic clam could survive the seemingly impossible journey through a water bird's digestive tract and emerge unharmed, as a chokecherry seed can pass through a robin. Or do young clams merely stick to the mud on a bird's feet while the flier travels to some other waterway? Some mollusks (but not this one) have indeed been found clinging to the lower feathers on a bird's body, as though they had clamped shut while the bird was bathing in shallow water.

The one explanation that seems unacceptable in the United States is that a few people keep introducing the edible clam where they expect later to collect some to eat. Only Orientals appear to have a taste for this delicacy. In the

Philippines, the little bivalve is lauded as a delectable morsel, a food that domestic animals (such as chickens) will eat,
and a good bait for anglers. If these uses for *Corbicula*
could be made widely popular in America, there would be
less reason for concern over the changing ecology of freshwater shallows, even if some two hundred small clams are
feeding and reproducing in every square yard. So efficiently does each individual filter the water in which it
lives, capturing the suspended green cells (such as diatoms)
as nourishment that a single Asiatic clam can clarify completely a pint of turbid water in less than two minutes. It
would be well to have such an animal on our side!

To some extent, the Asiatic clams undo part of the ecological damage inflicted by another oriental creature that is
now widespread in the fresh waters of North America. The
Asiatic carp was introduced deliberately about a century
ago by the U.S. commissioner of fisheries at that time,
Spencer F. Baird. He got the breeding stock from Europe
in 1876 and developed it from 345 fish to many thousands
in ponds near Baltimore and Washington. His agents promoted the carp as a highly desirable source of nutritious
protein, echoing the value judgments that had led Europeans to adopt the carp and its culture methods from the
Orient prior to A.D. 600. Apparently the monks of the Middle Ages bred the fish to mature earlier and perhaps produce more young than the carp that serve mankind in the
Far East. Carp culture came to England in the sixteenth
century, but cold weather prevented its spread farther
north.

Carp waste little energy on active movement. They gain
weight rapidly by scavenging in shallows for small plants
and animals. Standing almost on their heads, the carp suck
up the bottom debris, strain out the food, and discharge
mouthfuls of muddy water. Four barbels drooping below
the mouth provide keen sensitivity in touch and taste,
which permits the fish to feed where vision is impossible
because of the turbidity.

These habits cause no problems in a Chinese rice paddy, where the warm water between the emergent plants becomes a rich culture medium for microscopic algae. The green cells nourish newly hatched fish and also small crustaceans. The minnows grow larger and prevent outbreaks of mosquitoes, whereas the small crustaceans and any of the larger unattached water plants become food for carp. The fish add to the water ammonia, in the form of excretion, which acts as fertilizer for rice. Consequently, the yield of rice in paddies with a good number of carp exceeds that in fields without fish by 4 to 7 percent. Since hand labor is an essential part of oriental rice culture, workers wading about barefoot will notice if the minnow population gets large enough to interfere with production of carp. Peasants then "weed" the fish population with fine-meshed seines and feed the small fish they catch to ducks and pigs. The whole economy fits together smoothly.

Europeans combine carp culture with production of wetland crops other than rice. Ducks and geese dabble and cruise about feeding on tadpoles, preventing excessive growth of plants, and fertilizing the pond. Other fishes, such as tench and pike perch, can live in the same shallow water so long as their numbers do not exceed 10 percent of the carp population. On this regime, a carp will mature in about three years. A female will lay as many as two million eggs each spring; most of them may hatch in less than two weeks. The young fish can be five inches long by autumn, eight inches when one year old, and fifteen inches at two years. Kitchen scraps and other wet organic matter are added to the carp pond to enrich the water at low cost. The harvest in carp flesh seldom reaches the level characteristic of the Orient: twenty-seven to forty-five pounds to the acre of water surface, every two months during the growing season. But the yield is still impressive, and the nutritional quality of carp meat compares favorably with that of good beefsteak.

How could the respected commissioner of fisheries go

The Asiatic carp thrives in managed ponds of the Old World, producing protein for mankind, but it upsets the ecological community in streams elsewhere soon after it is introduced. (Photo by E. P. Haddon, U.S. Fish and Wildlife Service)

wrong with carp? First, he underestimated the resistance of Americans of European ancestry to the soft flesh of carp now that they had become used to the firmer flesh of cod, bass, trout, and other more vigorous fishes. Then he and his assistants overlooked one important feature of Eurasian carp culture: the confinement of the fish in ponds with no outlet, to be tended every few days. No warnings on this topic were issued along with the carp from the breeding ponds in the eastern states. By 1880 some fifty thousand carp had been distributed. Hundreds of thousands more were shipped annually for a few years to all parts of the country. Most of these fish were liberated in natural lakes and streams. There each carp followed its ancestral routine, rooting among the bottom vegetation, muddying the water if it could be muddied. The turbidity the carp produced blocked the penetration of sunlight and rapidly decreased plant growth. The fish themselves undermined the roots of cattails, causing these marsh plants to topple and die, dooming the marsh.

Most American fishermen agree that carp is a four-letter word and that catching these fish is an exercise in conservation rather than a pleasure. A carp, moreover, lives quite long (twenty-five years in the wild, forty-seven years in captivity) and has enough intelligence to learn from an early experience with a hook to avoid almost every bait. A later experience, when the carp is larger and heavier, seldom results in it being caught. With the hook in its mouth, the fish can swim off at top speed, with so much momentum that it snaps the line. Not until 1950, when the British angler Richard Walker developed special tackle and techniques, did a fisherman have much chance of taking a big carp in the wild. Now the number of people who have outfitted themselves and succeeded is growing—slowly.

But what do you do with a carp after you catch it? UN agencies urge that pond culture of these fish be expanded throughout the temperate and the tropical world, without explaining how a market for them can be generated. A superior variety of carp will yield annually as much as three hundred pounds of meat to an acre of water surface if little is done to improve the pond environment and up to four times this harvest if fertilization is intensive. Fertilization, of course, is an investment. It requires capital and presumes that the benefits will accrue to the investor. This is the type of operation that a rancher or farmer understands. The question of a market remains. The temptation is to begin a major program in a developed country rather than in a developing one (where capital is scarce), where the carp could be converted into fertilizer for land crops or into a salable form of fish meal as a nutritional supplement for poultry and livestock.

The hope of introducing an exotic fish that people will enjoy eating seems more realistic for the so-called mouthbreeders of Africa, freshwater members of the genus *Tilapia*. Africans have long used them for food. Fish fanciers occasionally maintain them in large aquaria, where the male of a pair in season will dig a small crater in the sediment and then entice a female to lay her eggs in the depression. He pushes her aside and adds his milt. Thereafter, one of the pair—the male in some species, the female in others—gathers the fertilized eggs into its mouth and broods them. *Mouth-brooders* would be a better name for them. After the eggs hatch and the young swim freely out of the parent's mouth, they will rush back in again at the slightest sign of danger.

Indonesians introduced one species of the mouth-breeder (*T. mossambica*) in 1939 and have since enjoyed this new food from small ponds on many islands. We met the fish first on the southernmost of the West Indies, where the Fisheries Department of Trinidad has some artificial ponds the size of swimming pools in which to raise *Tilapia* and test its appeal as an inexpensive protein. Trinidadians were buying all that could be produced at bargain prices. Six years later we found the fisheries officers in the Cape Province of South Africa embarked on a similar program with more than one kind of *Tilapia*. Their optimism proved contagious, particularly after we sampled some of the fresh fish cooked in various ways and found the flesh palatable and firm.

The parental care of eggs and young, which makes these African fish so intriguing, is apparently important in their home streams and lakes, where a great many kinds of predators abound. This behavior proves less important in the Western world, since *Tilapia* populations increase mostly where competition is minimal. Free in fresh waters, as they are in Florida, the introduced species become only minor members of the wild community. They do not alter the environment as the carp have done.

In a sane world, we expect people who hold authoritative positions to make themselves thoroughly familiar with the history of earlier gains and losses before committing a nation to programs for its future. "Move slowly and avoid repeated mistakes" seems the only policy. Yet every year live animals are freed deliberately in one country after an-

other before tests have time to show the probable consequences. A particularly frightening list of introductions is
published annually by the Food and Agricultural Organization (FAO) of the United Nations in its *Aquaculture Bulletin*
(formerly the *Fish Culture Bulletin*). The primary aim of increasing food for people seems often bypassed in favor of
other interests. Thus, the 1971 issue reports the placement
of a breeding stock of Asiatic carp from Malaysia in Fiji,
with the explanation that the Fijians need them as a source
of pituitary glands. We cannot imagine why. The following
year the *Bulletin* records the same kind of fish taken from
the United States to Cameroon, as though policies there
would avoid the problems produced elsewhere. The same
issue noted that carp in the Australian state of Victoria
make fresh waters so unsuited to other kinds of fish that
fishermen are supporting a program aimed at eliminating
the foreign species wherever its reproduction has been particularly successful.

The FAO tells also of freeing grass carp, a herbivorous
fish with coarse flesh from Asia, in some shallow lakes near
Stockholm, Sweden. The aim was not to provide a new
kind of food but to control an introduced waterweed. Grass
carp are on the banned list in the western United States
because of their history of undesirable effects on native
wildlife elsewhere—reducing the production of foods that
people like.

Policy decisions regarding fish often entail compromise.
One made recently in the northeastern United States is intended to affect both commercial and sports fishermen. It
reflects impatience with piecemeal programs that might
have restored the once-productive fishery for Atlantic
salmon. Formerly, these fish (which weigh as much as 102
pounds) spawned in the upper reaches of rivers from the
Delaware and the Hudson northward into Canada, as well
as along British and European coasts as far south as Portugal. Now the native salmon are gone from south of
northern Maine, seemingly unable to recover even though

some of the hazards of their environment are disappearing: harvesting has ceased, fish ladders have been built to let any salmon with the old instincts bypass many of the dams between the ocean and spawning sites upstream, and campaigns for pollution control have already brought measurable improvement in water quality. Yet only in Maine did federal and state agencies cooperate in restocking some of the smaller, least polluted streams, in the hope of restoring the fishery. Elsewhere the policy has been to introduce a different salmon from the West Coast.

Our Canadian neighbors have had greater success in reversing the downward trend in Atlantic salmon. By concentrating their efforts on the Saint John River in New Brunswick and releasing millions of young fish (smolts) annually from the hatcheries far upstream, they detect an encouraging increase in the wild breeding stock. Until recently, no one knew where the young fish went after they worked their way to the Bay of Fundy and into the Atlantic Ocean. But at least more came back. They followed ancestral habits that contrast markedly with the once-in-a-lifetime reproductive effort of the Pacific salmon, since the Atlantic species makes spawning runs year after year. The Canadian program might have gone farther toward restoring *Salmo salar* to its former abundance had not the fishermen of another nation found a way to interfere.

At first, the discovery of where Atlantic salmon go at sea was kept secret. Danish fishermen based in Greenland wanted the bonanza for themselves. Quietly they arranged to get suitable trawlers with which to harvest the edible fish that swarmed in such abundance in Davis Strait between Greenland and Baffin Island. Then their catch jumped from about 127 metric tons annually—less than 1 percent of the total annual catch of salmon from the North Atlantic and its rivers—to 1,539 metric tons in 1964. Early the following year, a reporter for *The Field*, a London news magazine, published a comment on the astonishing quantity of frozen Atlantic salmon on British and European markets, coming from Greenland by way of Denmark. Every year thereafter—except in 1968, when the trawlers got only 1,200 tons because drifting ice hampered operations—the harvest held high. But news of the source leaked out, and other nations sent exploratory expeditions. They found that 95 percent of the salmon in Davis Strait were young fish, weighing less than 7 pounds. Many of them bore tags that had been attached by hatchery personnel. Nearly 80 percent of the tags were of Canadian origin, another 15 percent were on fish that had left Maine as smolts, and the rest came from Great Britain and Ireland. This information caused a great outcry in Canada—"Save Our Salmon: Don't Buy Danish!" Magazines and newspapers in America carried articles entitled "New Form of Piracy" and "Danes Scourge the Seas."

An emergency meeting in London of scientific working parties from the International Council for Exploitation of the Seas and the International Commission for the Northwest Atlantic Fisheries gave representatives from Britain, Canada, and the United States a forum. The representatives pointed out that their nations were spending millions of dollars stocking the spawning streams in expectation of an economic return of fish weighing ten pounds or more. The Danes saw no relevance in this argument, since salmon chose to go to sea. The West German delegate, Gero Mocklinghoff, agreed that there is "no such thing as a national salmon." But when a vote was taken, only the Danes and West Germans favored a continuation of netting young salmon on the high seas.

The full impact of the Danish operations in 1969–71 would not be expected until 1974, when the fish that otherwise might have attained breeding age would be returning to their "home" streams to spawn. By then the Danish government reluctantly ratified an agreement that suited nobody: to terminate the high-seas harvest by 1976. Still angry, the Canadians wondered what else they could do prior to 1980, by which time conditions for Atlantic salmon might improve.

With such a background, it is no wonder that people in the eastern United States with an interest in fisheries looked for alternatives. They sought something better than carp, as a protein food that could be harvested from rivers and lakes. The landlocked Atlantic salmon in some fresh waters of New Hampshire and Maine appeared unsuitable, because they do not grow fast enough. In Lake Winnipesaukee, for example, the population can renew itself only because the catch each year is limited to sportsmen, between April 1 and September 30. Sportsmen are permitted to catch by trolling and to kill two fish (salmon or lake trout) daily, but only if the fish is at least fifteen inches long.

People wanted a new fish or a new method to raise Atlantic salmon profitably. They wanted the answer in the mid-1960s. Yet until late in 1971, no one in New England heard of the novel enterprise with Atlantic salmon that Johan Laerum had established six years earlier in Norway. His previous experience had combined a part interest in Mowi, Inc., a company producing marmalade and jam, and a hobby raising rainbow trout in seawater. When the business was sold in 1965, Laerum took over the company name and site. He invested in a steel-mesh fence set in concrete, closing off two acres of sea adjacent to his property. He also contracted for the use of a fish farm of similar size nearby. Into these two enclosures he released fifty thousand Atlantic salmon from a hatchery in Hardanger Fjord. This was close enough to Mowi, Inc., on Flogoykjolpo Island and the fish farm at Veloykjolpo, both about twenty miles southwest of Bergen.

The hatchling salmon grew in confinement, despite the fact that they had never seen a river. Within a year they became silver salmon, which is a step toward maturity not normally reached until between the end of the second and the fourth year of age. In two years more they became decidedly marketable adults, averaging about ten pounds. Laerum and his managing director estimated that their facilities were capable of producing annually about a hundred tons of high grade fish. They began planning toward 1975, when their scope might rise to five hundred tons annually and equal a third of the total catch of Atlantic salmon from spawning rivers around the world. Laerum just shrugs when he is asked, "Are these domesticated salmon? Are they tame?" At least his fish are easy to harvest, with no international complications.

In the United States, dreams of restoring a supply of delectable salmon to the river basins that open to the Atlantic Ocean turned in a new direction. The first idea was that of some Yankee fishermen in Connecticut, who began introducing a landlocked type of sockeye salmon (the so-called kokanee salmon, but still *Oncorhynchus nerka*) into East Lake, within the town limits of Salisbury. By 1965, anglers had caught a few mature males, which averaged 17.6 inches in length, and some females, averaging 16.1 inches. Although technically the fish was a red salmon and a real treat for the dining table, its flesh proved to be light colored and became even paler when cooked. Fishermen could not get used to the odd form and coloration of the head on the male, for it constitutes fully a third of the total length and includes a great hooked jaw with teeth .5 inch long. Moreover, the fish is blotched with violent red, often with a purplish cast.

In the West, most kokanee turn up in the smokehouse. In the East, a majority of these fish probably were eaten by brown trout, the population of which increased after the introduction. The venture proved interesting, but it scarcely seemed worth arranging for a new release of young kokanee every year to keep them present.

Then exciting news came from the Great Lakes, where a different attempt seemed to have a chance to start a new fishery. The old fishery formerly brought in about $3 mil-

Hunting for swimming prey in dark or turbid water, the sea lamprey employs pulses of self-produced electricity. This victim is a freshwater whitefish, to which the lamprey holds with a jawless sucking mouth while rasping flesh into particles fine enough to swallow along with blood. (Photo from *Life* magazine, courtesy of *Life* and U.S. Fish and Wildlife Service)

lion annually, at 1930 prices. But it ended during the mid-1930s with ecological disaster, because sea lampreys managed to reach Lake Huron and Lake Michigan. They destroyed the fishery resource of lake trout, yellow perch, walleye pike, and lake whitefish. Without these native predators, the alewives burgeoned. Nobody wanted them. They matured at a length of fifteen inches, although remote from the sea; bred; died; and were cast up by the millions to rot on lake beaches each summer, to the distress of vacationers.

A cooperative research program entered into by fisheries scientists from the United States and Canada finally found the weak point in the life history of sea lampreys. A specific chemical (a "lampreycide") was discovered. It would kill young lampreys in small streams where they grew for about eight years before entering lakes and attacking fishes of commercial size. Soon, everyone predicted, the lamprey population could be kept too low to matter. It was time to introduce an edible fish, preferably a gamey one, that would feast on young alewives. This could solve the alewife problem too.

The Michigan Conservation Commission tried steelhead

trout (*Salmo gairdneri*) from a West Coast hatchery. Those planted as fingerlings in 1965 delighted sportsmen as early as 1966. Why not add more steelhead and coho (silver) salmon (*Oncorhynchus kisutch*) as well. The salmon would grow bigger, if at all, and perhaps attain a weight of seven to ten pounds, as in Pacific waters. Early in the spring of 1966, the scientists released 50,000 steelhead and 350,000 cohos in Bear Creek (a tributary of the Manistee River), another 50,000 steelhead and 250,000 cohos in the Platte River of northwestern Lower Michigan, and still another 50,000 steelhead and 200,000 cohos in the Big Huron River of the Upper Peninsula. The released fish were all yearlings, the cohos about five inches long. A pound of them included about eighteen fish.

Five months later, on the morning of September 11, 1966, Gene Little fought for fifteen minutes before landing an 18-inch coho. It weighed more than 3 pounds. His excitement over this fish from the Big Manistee River led a farmer, Marshall Hoffman of Kaleva, Michigan, to admit that he had caught a 2.5-pound, 17.5-inch coho a few days earlier in Bear Creek. The news spread like wildfire among sportsmen. Taxidermy shops in Michigan began a roaring business. The boom in the tourist business continued, stimulated still more in the fall of 1967 when mature cohos weighing 15 to 20 pounds swam into the smaller streams to spawn and die. By then the Michigan Conservation Commission was aiming for the top. They had introduced 800,000 fingerlings of the largest kind of salmon in the world—the king or chinook (*Oncorhynchus tshawytscha*). By 1973 they hoped for giant fish, for those on the West Coast average 23 pounds when mature, and some individuals reach 80 pounds. It was worth introducing another 30 million yearling salmon annually if the lake would be productive.

New Englanders heard about the success with cohos and decided to imitate the program. The New Hampshire Fish and Game Commission released an initial lot of ninety

Mature salmon of several kinds make prodigious leaps to ascend waterfalls in North American rivers that flow into the Pacific Ocean, on a single, final trip to spawning grounds far upstream. Fisheries officers expect the fish to behave in a similar way to use the water cascading down a series of smaller steps in a fish ladder past man-made dams. Young salmon on their way to the sea may descend the ladder unharmed or be killed if they pass through the turbines of a hydroelectric generator at the dam. (Photo by G. B. Kelez, U.S. Fish and Wildlife Service)

thousand smolts that had hatched from a shipment of a hundred thousand "eyed" eggs obtained in Washington State. Fishermen watched with intense interest as the young fish began their slow movement downstream in two different tributaries, bound for Great Bay, the huge tidal area that connects with the ocean at Portsmouth. Everyone felt encouraged when the smolts reached seawater, already six to seven inches long, which is somewhat more than average for this species along the Pacific coast.

Some six hundred cohos of various sizes were caught between spring of 1969 and September of 1971, all of them near shore in the open ocean between southern Maine and northern Massachusetts. Would they know their way to their new home when they matured? Then adults began entering the fish ladder at Newmarket, New Hampshire, as though trying for a spawning run into the fresh waters of the Lamprey River, above a twenty-five-foot dam. A few more of these mature fish, with jaws a shocking shade of pink as a sign of their sexual condition, appeared below the dam on the Oyster River in Durham. We could see them clearly through the shallow water. Some were caught. Others died trying vainly to get past the dam. Each year maturing fish have repeated the attempt, a few in spring, more in autumn. It seemed a good time to repair the old dam and to install a new fish ladder, one that fish could use in the autumn of 1975.

Curiously, the prevalence of alewives and lampreys in Lake Ontario did not lead to any comparable attempt to change the situation. The New York Fish and Game Commission merely tried an introduction of young cohos in Spring Brook Reservoir, near Pulaski: 25,000 in 1968, 124,000 in 1969, 223,000 in 1970, and 122,000 in 1971. The survival rates averaged less than 1 percent. The few fish that did live long enough to grow in Lake Ontario grew poorly. Every one that returned to spawn had been attacked by lampreys and bore the characteristic circular wounds where the attacker had held on and liquefied some flesh to make a meal. Then, in 1971, the Canadians began using lampreycide. The New York commission asked the U.S. Fish and Wildlife Service to supervise a similar campaign in 1972. Only after the predator's reproduction stopped would it be worthwhile to send more salmon into the lake to feast on alewives. With luck, a growing fishery may be started by late 1976.

As we think about it, we realize how truly strange this whole new situation is. By rights, there should be no alewife population in any of the Great Lakes, nor any sea lampreys either. Both of these are properly marine creatures that come into fresh water only to spawn and spend their early stages in development. Both took advantage of man-made canals built to permit ships to bypass the fierce rapids of the upper Saint Lawrence River. Ships and fish began to move from tidewater near Montreal into Lake Ontario when the hydraulic locks were put into operation back in 1825, as the precursor of the present Saint Lawrence Seaway. Then lampreys and alewives were blocked from reaching Lake Erie by the 160-foot plunge of Niagara Falls and the surging cataract above the falls. These barriers were breached in 1829, when the Welland Ship Canal was opened. Within a few years, alewives appeared for the first time in Lake Erie. Fishermen there, and later on Lake Huron and Lake Michigan, welcomed this addition to their potential income. Alewives strained the plankton from the uppermost waters, where native fishes were few. Young alewives added to the food supply of predatory fishes deeper down and improved the commercial fisheries. It was not until the 1930s, when lampreys succeeded in getting beyond Lake Erie, that the situation got out of hand. Now the lampreys are being suppressed. The alewives, with ancestors that less than three centuries ago were in the Atlantic Ocean, have become food in fresh water for Pacific species of salmon and trout. Whether this new food relationship will maintain a state of uneasy balance may be discovered in less than a decade. Its weakest point is its dependence upon skillful control of lampreys and renewed generations of salmon in their new home.

Any ecosystem remains in peril if its future depends upon human management. The values that appeal to mankind fluctuate unpredictably, without much relevance to long-term benefits. Generally it is easier to transfer a living species to a new place on earth and to give it a generous start there than to direct the course of the human culture needed for its support. The classic examples include two successful ventures with fishes for New Zealand waters and two failures with introduced mammals.

Geological history made New Zealand an ideal site for a

simultaneous test of wildlife in relation to introduced animals and of human cultural responses. Until about A.D. 900 the two big islands retained a degree of isolation from the rest of the world that is hard to appreciate without an extended visit. The extinct moas, like the surviving kiwis and the few kinds of freshwater fishes, appear to have been leftovers from the breakup of Gondwanaland. This ancient supercontinent fragmented more than sixty million years ago into South America, Africa, Antarctica, and Australia. Somehow the ancestors of the flightless birds reached the islands of New Zealand after earth movements and volcanic action produced these sanctuaries, with a land area roughly the size of Colorado. Most of the other native birds seem to have come originally from New Caledonia and New Guinea to the north and northwest. New Zealand also attracted penguins and seals from Antarctic waters, petrels from all around the broad Pacific, waterfowl and bats from northward as far as the Siberian tundra, but few travelers from Australia across the Tasman Sea on the west.

The Polynesians who colonized New Zealand about A.D. 900 brought with them the domesticated dog and a peculiarly gentle Maori rat, a creature the men and boys made a game of hunting and brought home as food. Otherwise, the islands remained without land mammals. After 1840, Europeans established settlements and introduced a wide variety of cherished plants and exotic animals. Among these were the brown trout from Europe. The trout grew to huge size in the major lakes and rivers and reproduced well enough to keep sports fishermen delighted. Now trout can be caught legally day or night, and trout dinners are sold in restaurants close to popular lakes. King salmon from the Pacific coast of North America were tried. They encountered no serious competition and soon established their inherited pattern of maturing at sea and then spawning in their adopted New Zealand rivers. A profitable fishery developed, making food of the Northern Hemisphere available fresh to people in the Southern.

The European settlers of New Zealand met less success as they attempted to stock the islands with familiar or useful mammals. Today, only the domestic sheep, the tame cattle, and the European hedgehog are still regarded with approval. The wild animals with antlers or horns—deer and goats of various kinds—are now too small to interest sportsmen and so abundant that conservation officers must kill all they can merely to protect the native vegetation. The European rabbit, like the various predators (stoat, weasel, and fitch) introduced to reduce its devastation, finds few human friends and many foes. The Australian brush-tailed possum is still spreading and destroying valuable forest trees. The values that were anticipated from each of these introductions have been lost in the new cultural and ecological context. The mistakes prove not only expensive, but virtually impossible to correct.

Near the other end of the world, a similar plan went awry. At the time, the idea seemed brilliant. It was conceived by U.S. Coast Guard personnel who manned the lonely LORAN (LOng-RAnge Navigation) station on Saint Matthew Island, at the southern fringe of the Bering Sea. Occasionally parties of Aleut people would arrive for a short visit, usually when seabirds or seals were breeding on Saint Matthew, to make the trip there rewarding. For generations, the Aleuts had known how to reach Saint Matthew, yet none ever tried to live there, because nothing a person would want to eat grew on the island's tundra. The Coast Guard remedied that by importing twenty-four female reindeer and five fine males in 1944. This breeding stock was freed just weeks before orders came to abandon the LORAN station and hence leave the island once more uninhabited by mankind.

A reindeer herder from Lapland could have made a huge success from this enterprise. But Aleuts are not Lapps. Nor did they show the slightest interest in the food supply that now waited on Saint Matthew Island. The reindeer had ideal nutrition, with a great quantity of high-quality forage. Birthrates stayed high, and mortality low. When a Coast Guard party visited the island in 1957 and shot a few of the many reindeer, it found the males were 46 to 61 percent heavier than is usual for domesticated animals and the females from 24 to 53 percent heavier. Yet how many reindeer could a few coastguardsmen eat? The logbooks for all U.S. vessels that called at Saint Matthew between 1957 and 1963 show that a total of 95 animals were shot and used for food. Almost unmolested, the herd grew to an estimated 6,000 reindeer by the summer of 1963. Since the area of the island is only 128 square miles, this comes to 46.9 reindeer to the square mile.

The visiting party in 1963 noticed that body weights of the reindeer were down, even slightly below those of domesticated reindeer in well-managed herds. No longer could the animals find either their favorite lichen, "reindeer moss" (*Cladonia rangiferina*), nor the less-palatable worm lichen (*Thamnolia vermicularis*). Both had been destroyed by a combination of close grazing and shattering under so many hoofs. Grasses and sedges took the place of lichens. But these would be useless to reindeer in winter, leaving only the evergreen crowberry (*Empetrum nigrum*) in any quantity.

Winter came to Saint Matthew on schedule and provided more snow and cold weather than usual. Most of the reindeer died of starvation and exposure. The following summer, visitors found only forty-one females and one male alive. In large areas, the island was bare rock. To reduce competition among the scrawny survivors, ten females were shot. This management gesture came too late, for not one reindeer lived through the next winter. Saint Matthew is not only still uninhabited but more uninhabitable than ever because its former tundra plants are mostly gone.

We scarcely can blame the Coast Guard personnel for their attempt to help the Aleuts. Yet Saint Matthew Island might have fared better had these officers known that

Aleut culture is essentially the same as that of the Eskimo—their nearest kin—and that the U.S. government, through its Bureau of Indian Affairs, had already invested about $3 million in trying to interest Eskimo in herding reindeer, with no discernible success. This expensive program began in the 1890s when five hundred reindeer were purchased from Lapland and three young Lapps were hired to herd the animals from Siberia across the frozen Bering Strait to Alaska. But as fast as the government gave the reindeer to Eskimo, these northern people slaughtered them for a feast, turned them loose to mix with native caribou, or sold them.

Bureau officials persisted in their give-away policy and kept the captive herd healthy and increasing by appointing more Lapp herders to handle the animals. This was the situation in 1914, when Carl Lomen and his two brothers arrived in Alaska, looking for a way to make money. They chose to buy reindeer from the Eskimo at every opportunity and to pay Eskimo laborers to do what no Eskimo seemed willing to do for himself—to take care of a growing herd. By the 1930s the Lomen herd totaled 250,000 head. The rate of slaughter was adjusted to match the rate of reproduction. It supported a meat-and-fur business that had no equal in the Far North, but benefited only the few Eskimo that worked for wages.

The Bureau of Indian Affairs sought a bureaucratic solution to this situation by getting Congress to pass the Reindeer Act in 1937. It made illegal in Alaska the possession of a reindeer by anyone other than an Eskimo. The territorial government then bought the Lomen herd at $6.50 a head and tried once more to give the animals to Eskimo. Out of a tribal population of 22,300, fewer than a score were found who would accept the herding task. One of them was Johnson Stalker, who now manages Alaska's first model reindeer farm under supervision of the University of Alaska, on a half-million acres of government-owned land six miles west of Nome.

Much of the expertise for handling this herd comes from John Zumstein, a livestock farmer of Redmond, Oregon. His interest in reindeer led to his acquisition of some government-surplus animals in Alaska and taking them to Oregon, where the Reindeer Act would not hamper his enterprise. In his first fourteen years of operation he built up his breeding stock from twelve animals to more than two hundred, all of them stronger, heavier, healthier reindeer than those with which he had started out. Zumstein's methods, based on "horse sense and a lot of curiosity," plus the dedication of a hobbyist, may yet give Alaska an industry based on renewable agricultural resources. The critical step lies in incorporating into Eskimo culture a suitable domestic animal of any kind.

No doubt crises of this sort have been passed many times during the six-thousand-year history of raising animals and plants for human benefit. What seems new is a humanitarian concern shown by the dominant population in North America over the threatened minority in the Far North. Those whose heritage encompasses the beneficial use of domesticated kinds of life feel reluctant to have their advantage cause the disappearance of the Eskimo. The nearest parallel is in Australia, where the Caucasian settlers try to accommodate the aboriginal tribesmen, who decline to raise useful plants and animals. No such humanitarian generosity has been at work to save the Pygmies (whether Negrillos in Africa or Negritos of Asia), who once occupied sizable territories, or the Bushmen, who, in their day, held an even larger part of southern Africa.

The truly strange aspect of a dependence on domesticated plants and animals for food and fiber is the cultural resistance to augment the list of those that have already been adopted. Much of the ecological distress in the modern world could have been avoided—and to some extent still can be—through introduction by the peoples of a continent of only those types of life that are already well adapted to the climate of that continent. Europeans could well have made a domestic use of red deer instead of importing sheep from Asia Minor. As colonists in North America, they had an opportunity to tame the caribou, to raise bison and pronghorns as a reliable source of meat and hides. South of the great deserts of Africa, the eland and other antelope offer a comparable opportunity, producing more food per acre than is possible with cattle, sheep, or goats. Australian settlers could have turned to domesticating the mobs of kangaroos, instead of introducing cattle, sheep, and—in their tropical northlands—water buffalo, which now run wild. Each of these alternatives has been tested recently and found to be eminently suitable. Only our cultural inertia stands in the way of a gentler compromise between mankind and the living things that are native to lands where our populations live.

5

Targets and Trophies

SECOND LARGEST of the lovely islands in Puget Sound is San Juan. Although part of Washington State, it lies closer to Canada's Vancouver Island. We took a ferry there recently, to visit the famous oceanographic laboratory operated by the University of Washington and to see some European rabbits that are feral on American soil.

The rabbits make a lasting impression, for they keep the ecology of the island out of joint. At the southeastern end of San Juan they maintain a major warren. Their interconnected burrows honeycomb the fields and slopes. Almost no vegetation remains there, except stunted bracken, Canada thistle, cheat grass, and tarweed. By day both golden eagles and bald eagles soar silently above, watching alertly. Yet these swift birds of prey catch only a few unwary rabbits or sick ones, for the European rabbit is naturally nocturnal.

The ancestors of the San Juan rabbits were domesticated stock, imported and released about 1900 by the lighthouse man. He hoped to enjoy rabbit meat as food that would cost him only the trouble of catching the animals. By 1924, his successor at the lighthouse had begun to beg the U.S. government for help in controlling the rabbits. None came,

and the rabbits continued to multiply. Some of the same
breeding stock was released on smaller Smith Island
nearby. There the situation seems beyond rescue. Buildings
have been abandoned because their foundations were undermined by the rabbits, and the coastal bluffs continue to
collapse, shrinking the land area. The animals number
about thirty per acre and have no future that anyone can
see.

One man on San Juan itself long made a living out of
catching rabbits. With a jeep and a long-handled scoop net,
he captured them at night. He claims to have shipped out
more than fifty thousand annually for many years. Now his
methods have been refined for visiting hunters. They come
to three resorts that advertise bed, board, and bunny nets.
Available are rental cottages, hookups for travel trailers,
and campsites, all at relatively modest prices. The profit to
the islanders comes from the charter use of converted automobiles with special seats, open bodies, powerful headlights, and long-handled nets. Four people can reserve such
a vehicle for less than $10 each, and use it for two hours or
until they capture eighty rabbits—whichever comes first.

We talked to one of the men who offers this service. He
assured us that it is unusual for four people to need two
hours to catch eighty rabbits. The animals are abroad in the
darkness, searching for food. They freeze when headlights
approach and can be captured easily. From net to gun-
nysack takes only a minute.

"What can anyone do with eighty rabbits?" we asked.

The man grinned. "Most people just empty out the bags
and let the rabbits go, to give someone else a good time. It's
about two hundred pounds of rabbits. But if hunters ask
me to, I'll dress the carcasses at two for a dollar and pile up
the skins. They're not worth much."

The deer that the islanders advertise are not worth much
either. Those we saw were several sizes bigger than the rabbits, but abnormally small for deer. In fact, we mistook the
first ones we noticed for fawns that had prematurely lost

their pale spots. Later we encountered deer of the same size with smaller, fully spotted fawns and realized that unspotted ones are mature animals. They cannot compete effectively with rabbits in clearings or along the forest edges. And in the dense woodlands, neither a deer nor a rabbit can find much to eat.

The rabbits make farming on San Juan difficult. A cleared pasture is merely an invitation to the first pregnant rabbit that arrives to dig a burrow and start a new warren. Yet some islanders persist. We did see some grainfields and a few pastures with either black-faced sheep or white-faced cattle. Raising these animals requires ingenuity and expense in poisoning rabbits without endangering livestock or pets. But the poisoned bait, the traps, the dogs and cats, the wild raccoons and mink, all combined cannot compete with rabbit reproduction. The only real control over San Juan rabbits after three-quarters of a century is a habitat that the hopping hordes still find unsuited to their needs.

On American soil, the introduced rabbits have become live toys to pounce on in a nocturnal game that sounds more hazardous than it actually is. The game sustains the illusion of dashing about after an elusive quarry and of capturing something that could be eaten. It qualifies as an outdoor sport, an amusement, a diversion, a shared experience to be relived afterward. If no rabbit is killed and no blood is spilled, the game seems harmless. The attendant is well paid for the upkeep on his noisy vehicle. Other islanders benefit from the spending by the visitors. Still others seek some middle ground. Either they erect large signs inviting Hunting By Permission (which is free to sober people with suitable equipment) or Rabbit Hunting (which is for a fee, and supervised). Or they resent the nightly noise of guns and fear a misdirected bullet enough to post No Hunting or more specifically Don't Shoot—Children At Play, even if the youngsters are regularly in bed before the fireworks begin. In either direction, the presence of European rabbits limits the islanders in the diversity of values their land can offer.

The idea that a rabbit might be netted or shot on mere whim affects people differently. One person who "enjoyed the gentle, soft-spoken little bunnies who hopped harmlessly beside the roads and through the brushlands and held no malice for anyone" on San Juan Island went on to object, as a casual observer "neither emotionally nor economically involved," that these animals "were here long before we were." Apparently this visitor, who had spent some years on the islands in Puget Sound, did not learn the rabbits' history; neither did he relate the nubbly carpet of rabbit droppings to the devastated land at the one end of San Juan or visit Smith Island, where the relationship is unmistakable.

The European rabbit has been misunderstood many times before in the last two thousand years. It may not even be a European animal, for the only native populations known to mesh smoothly with their environment are those of arid lands in North Africa. Quite possibly the rabbit originated somewhere between Morocco and western Tunisia. It may have entered southern Europe as a semidomesticated possession of Phoenician traders about 750 B.C. or of adventuring Moors about A.D. 750. Almost certainly the Normans, who liked to hunt and eat rabbits in France, missed the animal in England and introduced it there soon after 1066. The progeny dug in securely.

Englishmen not only developed a fondness for rabbit shooting but selected from among the natural diversity of domestic dogs until they had a beagle race that would excel in rabbit hunting. It was an Englishman, Thomas Austin, who sought to improve his estate near Geelong in the Australian province of Victoria by adding European rabbits in 1859. He ordered two dozen from home and freed them on Australian soil just four days after they arrived aboard the Black Ball clipper *Lightning*. Later, Austin claimed that he and his guests shot 20,000 rabbits on his estate in the first six years, which averages 9 rabbits every day. He estimated that another 10,000 remained as breeding stock. Still later, in this same area, the rabbits demonstrated that even if two-thirds of them are shot every year after the first, 24 rabbits could become 80,368 in three years and 22 million in six. Each pregnant female rabbit (doe) can produce up to 10 litters of at least 6 young annually. A harvest of only 20,000 would barely nick the population!

Neither Australia nor New Zealand had any native predators to hold back the rabbits. Some house cats went wild, but Down Under they found the native marsupials easier prey than rabbits. The Australians introduced foxes from Europe, only to discover that the foxes preferred the small, plump, rabbit-sized rat-kangaroos known as tungoos. Tungoos and rabbits came to live together amicably, the former often occupying the mazes of tunnels that the rabbits dug in each warren. Tungoos, however, reproduce slowly. With foxes for competition, they soon became a vanishing species.

The feral house cats increased in numbers and caught so many of Australia's unwary birds that insects suddenly prospered because of relief from their flying predators. Leaf-eaters and boring insects multiplied, attacking the native eucalyptus trees with unprecedented vigor. Among the bare branches the human colonists often saw koalas exposed, each of these teddy bears trying to find foliage the insects had missed. To save the eucalyptus forests, the Australians began exterminating the koalas, shooting these slow-moving creatures wherever any could be found.

European rabbits spread north and west from Victoria until they occupied more than a third of Australia in spite of hunting, poisons, traps, dynamited warrens, introduced ferrets, foxes, dogs, and a supposedly rabbit-proof fence of wire netting stretched more than seven thousand miles across the country. Australians who profited from catching rabbits were known to drive to the expensive fence with a truckload of captured animals and toss them over the barrier, to run free and reproduce on the other side. Much of the profit came from the export of 700 million rabbit skins and 157 million frozen rabbit carcasses in a single decade. This supplied most of the world's demand for felt from rabbit fur and the ingredients for countless rabbit pies and rabbit stews. If the rabbit hunters had fun and made money, nearly everyone else lost. Five rabbits ate as much as one sheep. The combined payments for 5 rabbit skins and 5 carcasses amounted to less than a third of the value in one sheep's annual crop of wool, ignoring all gain from lamb or mutton, hides and tallow.

These conflicts in interest did not end after 1950, when microbiologists and wildlife managers for the Australian government began field tests with the virus of myxomatosis. Citizens in the rabbit-catching industry gloated quietly over the failure of the experiment and then suddenly felt threatened in December when rabbits began dying by the thousands all over the Murray-Darling drainage. At better than three miles each day, the disease spread up the river valleys in every direction. Everywhere, it seemed, the rabbits would soon sicken and become sleepy, while their skins would break out in great sores. Mosquitoes and other blood-sucking flies visited those sores and got myxoma virus smeared over their mouth parts; then these parasites flew to healthy rabbits and inoculated them in turn. The only objections came from animal lovers who were offended by the sight of dying rabbits and from the hunters, skinners, and meat-packers who relied upon rabbits for a livelihood. These people took heart in March 1951 to see some surviving rabbits hopping about as though nothing had happened. Sportsmen ignored the whole operation, for they had long ago become bored with rabbits as a test of marksmanship.

When the mosquitoes began flying in the Australian summer of 1951–52, they encountered newly inoculated, freshly released European rabbits in practically every infested part of the country. Never before had a deadly disease swept through an animal population with such devastating effect. Out of every 1,000 rabbits in Australia, at least 995 succumbed to myxomatosis. Often the survivors ran about wildly by day, as though seeking company, and met a fox or a farmer with a gun. Grass sprang up where none had been seen within human memory. Sheep put on weight almost twice as rapidly as before, and pastures supported far larger flocks. Australian landowners began to dream of a continent without rabbits. But members of the rabbit industry had dreams too—of obtaining and using a vaccine that would immunize rabbits against myxoma virus. Over the next two decades they tried to save their enterprise, although they knew that this was sabotage. What they could not predict was the new determination among their countrymen to rid Australia of every single rabbit.

New Zealanders, who normally avoid noticing (let alone imitating) what happens in Australia, could not ignore the glowing reports on myxomatosis in the neighboring country. But, for some reason, attempts to duplicate the success on New Zealand soil led nowhere. Yet so aware of their environment are most people in the two big islands that they encouraged their government to move in other directions. Legislation wiped out the traffic in rabbit skins and meat by making it illegal. Antirabbit drives, including the spreading of poisons by airplanes in mountainous and unpopulated areas, were stepped up. Even more important was the progressive transfer of responsibility for exterminating rabbits from individual landowners to locally elected "rabbit boards," toward which both the landholder and the government contributed tax funds. Every rabbit became an authorized target.

Members of the rabbit boards discovered in one area after another that the prime problem with rabbits was not rabbits at all but depleted, eroded soil, on which they thrive. Warrens spread where too many sheep and cattle use the land and keep it almost bare. Drought also befriends the rabbit, and a good grass fire helps it considerably. Every measure improving the soil, even reseeding it with good grasses, reduces the number of rabbits. If the land is fallowed—used for grazing only on a rotational basis—rabbit warrens disappear. By 1958 the New Zealanders could boast of great progress, even though myxomatosis had been no help to them.

We found members of rabbits boards searching in 1966 for other ways in which they could improve their environment. Rabbits had virtually disappeared. Now we wonder what a few experienced members of such a board could do on the islands in Puget Sound. We could enthusiastically endorse that kind of missionary effort. It would be a novel Peace Corps endeavor among members of the English-speaking world and could help restore nature from a seriously unbalanced state. Myxomatosis and rabbit fleas to carry the disease might do wonders on San Juan.

The New Zealanders have many another challenge from mammals that were originally introduced as targets and potential trophies. What a worldwide collection they face, all established in the name of sport:

Red deer (*Cervus elaphus*) from Europe, on South Island since 1851

Fallow deer (*C. dama*) from Europe, on both islands since 1864

Axis deer (*C. axis*) from India, on both islands since 1867

Eurasian hare (*Lepus europaeus*) from England, on both islands since 1867

Wapiti (*C. canadensis*) from Canada, on South Island since 1870

Wallabies (several species) from Australia, on South Island since 1870

Sambar deer (*C. unicolor*) from India, on North Island since 1875

Japanese deer (*C. nippon*) from Manchuria, on North Island since 1885

Chamois (*Rupicapra rupicapra*) from Austria, on mountains of South Island since 1889

Moose (*Alces americanus*) from Canada, on South Island since 1900

White-tailed deer (*Odocoileus virginianus*) from the U.S.A., on South Island since 1901

Tahr (*Hemitragus jemlaicus*) from Himalayan Asia, on mountain slopes of South Island since 1904

Every continent except Africa and South America is represented. Nor were these two boycotted. Attempts to start herds of zebras and gnu failed, as did those of llamas and alpacas. But antlered members of the deer family and the goatlike chamois and tahr have succeeded too well. In a suitable habitat, overpopulation has become normal; growth is regularly stunted. Sportsmen see no trophies and few targets worth hunting. The government, which sponsored these introductions, now not only authorizes hunters to kill as many of these animals as they wish, on any day or night, but encourages them with free ammunition. Even this provides so little control that wildlife officers must go out periodically with helicopters and automatic weapons to reduce the half-starved populations. The introduced animals have ceased to be game or to provide real sport. The patriotic goal now is to eliminate them and thereby protect the land.

When we chat with active hunters in North America, we find most of them unconvinced that anything other than gross incompetence by wildlife management programs could let a game animal become an official pest. Without experience in the foreign context, these Americans assume that trophy-hunters would have prevented so uncontrolled a rise in target populations, if only the laws had permitted the voluntary operation.

These same men and women, we notice, criticize the policy wherever hunting is prohibited and wildlife managers must control large grazing and foraging animals repeatedly to preserve the habitat. This need arises whether management is biased toward the welfare of all species or a single one, such as on the U.S. National Bison Range near Moise, Montana. The convictions of hunters arise from a mystique that equates the impact of modern sportsmen with the natural influence of major wild predators—animals that are no more. A real change in this tradition cannot be expected until people understand more fully the role of the predator and the long-term genetic consequences of hastening death selectively. A predator conserves its energy by attacking only the most handicapped prey it can find. A hunter chooses a healthy animal in the near-trophy class or just any animal if frustration rises high enough. We have no analogy in a human population, for who could gain at the expense of the unwary and unwanted (such as Skid Row derelicts) rather than leaders and workers of the community (who are worth a ransom)?

At the National Bison Range and in national parks from Canada to South Africa, we see wildlife managers who do appreciate the need to favor animals in the target or trophy class to perpetuate their kind. When a reduction in the number of any species seems essential to safeguard the food supply and space for the rest of the population, these people choose individuals as a predator might. They take out the weakest young, the animals past their prime, the victims of accidents that survive only because they are in a protected environment. And so far as possible, the meat and hides from each annual culling go to reduce the cost of some tax-supported enterprise. "Useless" parts return to the soil in much the same way that the leavings of predators would.

When we think about it, we realize that modern wildlife managers have learned through science what the old gamekeepers knew and practiced on big estates in Europe in centuries past. Then game was a perquisite of affluent nobility as well as a product of a tended habitat within a fenced perimeter. The gamekeeper kept close track of every deer and could tell when a doe grew old or a buck's new pair of antlers was slightly smaller than those he shed last year—a sure sign that a buck is past his prime. The owner of the estate understood why his gamekeeper would drive the healthiest animals into a separate paddock before sportsmen friends were due to arrive for a scheduled hunt elsewhere on the grounds. Those the sportsmen did not shoot on the gala occasion the gamekeeper might harvest for the larder of the estate's big house. The intent, after all, was to maintain the quality of both the game animals and the preserve on which they lived.

We can only wonder how far back in history this pattern of affluent sportsmen and their ritualized hunting might be traced. Surely at some time, the killing of meat animals for reasons other than hunger must merge with the acquisition of food either routinely or through real urgency. Possibly the origins lie less in the appearance of an affluent minority than in claims to special rights and rites by the first priesthood. Diversion of food to appease supernatural gods may well have been the earliest luxury of Stone Age people, as a token investment toward a hoped-for benefaction.

Perhaps there was an earlier stage, when the usefulness of the trophy remained secondary to the achievement of securing it. When small boys first learned to throw stones and sticks with greater accuracy of aim than any modern chimpanzee, the target must almost certainly have been some active bird or mammal. We may ask if culture was then far enough advanced for a youth to shout in some language, "I hit it! I killed it!" We have seen this deadly behavior practiced at the expense of small birds that settle for a moment on the African plains near a boy who is herding the livestock. Whether the carcass goes into the soup pot or not has little consequence. The boy has proved his skill. At a step removed, this demonstration has its equal in the pride of a Masai youth who has killed a lion single-handedly, armed only with his spear and leather shield. He has earned a place among mature tribesmen by risking his life in this way and winning. To the disappointment of Masai youngsters, this tradition has had to be abandoned within the last few decades. No longer can these boys dream realistically of an opportunity to have their day of glory and from it a lion's mane and some of its fur to wear for personal adornment. The Masai population has increased until the number of young men needing to prove their manhood exceeds the number of lions in the country.

The disappearance of wild animals, whether exploited for trophies or food or both, has followed each rise in human population for many millennia. One of the most precipitous declines occurred in North America just after Paleolithic hunters arrived in northwestern Canada and began to spread south and east. In a mere five hundred years, ending about eleven thousand years ago, about two-thirds of the continent's large mammals became extinct. The seventy-odd species that vanished so quickly included mammoths, mastodons, wild horses, camels, several kinds of bison, ground sloths, saber-toothed cats, giant armadillos, and large rodents. Since their ancestors survived all the challenges of the Ice Age, including its potential for "overchill," something new must have happened to turn a former abundance into zero populations.

Paul S. Martin of the University of Arizona and other geoscientists since 1967 have used computer simulations to examine the possibilities of "overkill." How could animals that were thriving well enough to have left many good skeletal remains as fossils vanish so abruptly at human hands unless people were far more numerous and well equipped than seems apparent from the remains of their scattered campfires and their few artifacts? But calculations indicate that the known extinctions could have been accomplished unwittingly in less than three hundred years by a hundred hunters of both sexes and their descendants.

Before the arrival of people, the population of large game animals was presumably between the average for herds on African game parks today and the present average in the United States for all livestock and wild game as large as, or larger than, a white-tailed deer. Suppose one person in each family of four did most of the killing and felled an animal weighing one thousand pounds (an "animal unit") each week. Not more than half of the meat would be eaten, but the rest would spoil or be consumed by scavengers while people slept. To obtain such a frontier diet, averaging ten pounds of meat per person daily, the tribal group would have to advance as much as twenty miles each year if the human population exceeded two to the square mile. By the time the advancing front of this hunting population reached the Atlantic coast, the Gulf of Mexico, and the northern edge of the Mexican plateau, close to one hundred million animal units would have been destroyed on some three million square miles. The total number of people engaged in this spectacular history, sparse colonization, and extermination may have been less than a third of a million. If life expectancy then was only twenty-five years, no more than one hundred thousand may have been alive at any one time. So few people over so large an area would be almost invisible in the fossil record.

Martin and his colleagues could not resist using their computer to explore one idea that has become popular recently. Suppose that these hunting people had developed a cultural taboo, one that kept them spread out with no more than an average of 1.95 people per square mile. What if they had introduced a program of zero population growth as soon as they could find no more territory to colonize? These traditions would prevent a tribal group from forming a "front" and causing extinctions. People would have occupied North America in as little as 330 years. Thereafter, with a population stabilized at just under 6 million, they could have continued to eat ten pounds of meat apiece daily for an indefinite future. Such provident hunters would have left a truly astonishing array of animals and a wealth of artifacts to delight modern archaeologists.

Instead, the hunters seriously diminished the resources of the continent. Even when later cultures developed a reliance upon domesticated strains of maize and other plant foods instead of meat, the Amerindians north of the Rio Grande increased to no more than 1.2 million. This

number sparsely occupied the United States and Canada in
A.D. 1620, according to the most reliable estimates. The Indians averaged no more than 1.7 to the square mile. Neither the total for the populations of Amerindians in these
two countries nor their level of subsistence is much different today.

We can think of the pronghorn "antelopes"—the most
fleet of North American mammals—and the sole surviving
species of bison ("buffalo") on the continent as having inherited the heartland. Neither suffered unduly from Indians until firearms were introduced. Even then, the rare
white bison were safe. No Indian would hasten the death of
these seemingly supernatural animals; only a chief might
wear a robe made from a white bison that died of old age.
Nor did the Indians interfere seriously with the magnificent elk or the smaller members of the deer family, which
maintained their populations so long as they had to face
only Indians without guns, mountain lions, grizzly bears,

The pronghorn is North America's most distinctive mammal. On
no other continent does a cud-chewer shed each year an outer
sheath from a bony core of horn. Highly adapted for life on semiarid plains and foothills, the pronghorn races away from danger,
chiefly into the unpeopled fringes of former bison territory.
(Photo by E. P. Haddon, U.S. Fish and Wildlife Service)

wolves, and lesser predators. Moose roamed the swamps farther north. All of these animals found areas where Indians rarely went, either because the traveling proved arduous or because it was a sort of no-man's-land between tribes. Any Indian hunting party on intervening land was likely to be ambushed by other, hostile Indians from the other side, perpetuating ancient feuds. Game animals were safe there except during those rare periods when local warring tribes established a temporary truce. Then hunters might enter the buffer zone freely in search of deer, elk, moose, or bears.

Not until the last century were the pronghorns fenced away into the desert fringes and the foothills of western mountains. Fences provided a secure place for cattle and sheep from the Old World but a barrier that a pronghorn would not jump.

Neither a bull-proof fence nor a muzzle-loading musket interfered much with the continent's grizzly bears. C. Hart Merriam, who directed the old U.S. Biological Survey, which was the forerunner of the Fish and Wildlife Service, recognized eighty-six different races of this largest carnivore on earth, from those in Alaska to the Mexican population. By 1939, a census of grizzlies south of Canada revealed only 1,100 individuals, none of them in California, although that state continues to show the big brown bear on its official flag. In 1949, Montana had 570 grizzlies, Wyoming 120, Idaho 50, and Washington 15. The Mexican bears dwindled until 1970, when the last of them succumbed to rifle fire. Today Alaska and western Canada have most of the surviving grizzlies. Only a few remain in Glacier National Park (Montana) and Yellowstone National Park (Wyoming). Should they too be "removed" to make the parks safe for people?

Hunters and salmon fishermen would like to kill the grizzlies that come each spring to Alaska's salmon streams. The powerful bears pounce upon full-sized fish and drag them from the water, sometimes for only a bite or two of

Grizzly bears have lost to mankind all of their former range south of the Canadian West except for small areas in two national parks. Even there their right to roam and to defend their young is contested. In Alaska, where grizzlies attain record size as the world's largest surviving carnivores, salmon fishermen covet the fish the bears scoop out with dexterous power when the annual spawning runs occur. (Photo courtesy U.S. National Park Service)

flesh. But a recent investigation by George W. Frame reveals that this "loss" of thousands of salmon does the fishery more good than harm. Only about 8 percent of the fish the grizzlies catch are still loaded with eggs or milt, on their way to spawning shallows. The rest are expended fish, destined by their genetic heritage to die without returning to the sea. Many of them already have fungus on their gills. By hauling out these salmon, the bears interfere with the spread of the fungus to still-healthy fish and keep thousands of fish carcasses from polluting the water. This conserves the dissolved oxygen, leaving it available for the growth of salmon embryos and hatchlings. The grizzlies perform a service few have appreciated from a safe viewing distance.

America's vast herds of bison could not withstand indiscriminate slaughter. From perhaps 50 million animals (some weighing 3,100 pounds) in 1801, only 23 dark-haired individuals remained alive in 1901. They were given sanctuary in Yellowstone National Park where the herd built up. It reached the carrying capacity of these particular 3,864 square miles. An allowable total of 700 bison had to be agreed upon. Excess animals would be shared with other parks and zoos or sold to ranchers who wished to raise bison.

A separate herd of these native animals was established in the National Bison Range near Moise, Montana. Among its members a white bison was born, a blue-eyed male that became a dominant individual, siring a host of dark-haired offspring. Eventually he died of old age. Named Big Medicine, he is a display specimen in the state museum in Helena.

Before their relocation to Yellowstone Park, the bison had contracted from range cattle a low-grade infection of Bang's disease (brucellosis). The bacterial parasites cause no harmful effects in bison. But now cattlemen are seeking to have America's bison eliminated, because they fear that brucellosis will spread even across fenced buffer zones to valuable cattle, in which the disease causes spontaneous abortion. The native animals are again in jeopardy, regarded as a continuing menace to the livestock industry.

The value placed upon wild animals undergoes continual change. At the beginning of the twentieth century in North America, it reflected a new sense of freedom among people whose ancestors had chafed against conservative customs in Europe. By 1901, the American people established some novel forms of land use to match their chosen values. Under a flag with stars to represent forty-six states, the U.S. Department of the Interior took responsibility for Indians, fish and wildlife, national parks, public lands, mining, and a detailed survey of geological features. Almost 200,000 square miles were allocated to Indians, on 118 tribal reservations. More than 6,000 square miles were allotted for six national parks. Nearly 1.5 million square miles remained federal land. A slightly larger area, owned

"Big Medicine," the white bison with blue eyes that was born, became dominant, and died on the National Bison Range in Montana among the herd that is protected for study there, still stands on display in the state museum. In former times, an Indian chief would have claimed the right to wear as a ceremonial robe the skin of so rare a bison. (Photo by E. P. Haddon, U.S. Fish and Wildlife Service)

privately, served the needs of about 75 million people of Old World origin. Close to 40 percent of them engaged in making the cities grow.

In 1901, rural areas had an average of only twenty-eight people to the square mile. They produced enough food and fibers from introduced animals and plants to supply the city-dwellers and markets overseas. Among these claimants to the continent were many with the time, personal interest, and financial means to hunt wild animals as targets and trophies. Unlike the Indians, these people chose their quarries less for food to be gained at small cost than for challenges to be met in making a kill. Was the target elusive, like a squirrel or a dove? Did it excel in hiding, like a deer or a pheasant? Or, following English usage, was it "vermin" because it sometimes preyed upon game animals or domestic livestock? Sometimes the motive for slaughter was simply proof of the boast "My gun is better than your gun."

Today, trophies are few. Elk are gone from most of their original range, unless they have been reintroduced. Only the male elk offers a tempting head, with a spreading rack of antlers. Both sexes possess two teeth deemed suitable to wear as pendants on a watch chain. Now targets tend to be deer, the white-tailed in the East or the mule deer in the West; ducks among migratory waterfowl; introduced pheasants; trout, among freshwater fishes; or smelt for those who enjoy fishing while ice covers the estuaries and larger lakes.

Deer are creatures of the forest edge. Its shrubby growth affords them the browse they relish. Each logging road, clearing, or right of way increases the total length of edge and the opportunities for deer. So long as the deer have hiding places for the day in patches of woodland, they multiply because of human activities. Often they reproduce so prolifically that more of them seek food in late winter than there is browse available. Many starve to death or succumb to diseases that become lethal in malnourished animals.

At dawn or dusk, we often see a doe or an antlerless buck. In hunting season both sexes hide away. Despite their numbers, these animals are seldom noticed except by sharp-eyed people. Hunters overlook their chosen trophies by staying close to roads (which saves the labor of carrying a dead deer through brush to a parked car), whereas the deer spend hunting season as far from roads as possible. Between hunting seasons, deer cross roads so freely that, despite signs warning Deer Crossing, they collide with passing automobiles to the detriment of both deer and vehicle. While snow is deep, deer often travel along the packed trails left by snowmobiles. Outdoorsmen may collide with a deer inadvertently or encounter it in some place where rules defining the hunting season and mode of kill can be ignored with little risk of detection and prosecution.

Fences built to restrict the wandering of domestic livestock now limit the natural migration pattern of native elk (wapiti) wherever they survive or have been reintroduced in the United States. Those in the Jackson Hole country of Wyoming have adjusted to spending the winter on a national refuge where supplementary food is made available. In spring the elk move out of the refuge, up the mountain slopes where the young are born and a lush growth of new vegetation supplies needed nourishment. (Photos by E. P. Haddon and C. J. Henry, U.S. Fish and Wildlife Service)

From southwestern Canada to Mexico, mule deer transform browse into more of these agile animals and thereby furnish food for western cougars and targets for human hunters. Annually the hunters take about a tenth of the mule-deer population, the combined action of cougars, bears, coyotes, bobcats, and lynxes, perhaps another tenth. The deer population remains stable only if starvation and disease account for fully another tenth, since the animals themselves show no ability to regulate their own numbers. (Photo by E. P. Haddon, U.S. Fish and Wildlife Service)

Migratory waterfowl receive more complex and skilled management. The U.S. Fish and Wildlife Service compiles an annual census for each species. This information is shared with state fish-and-game commissions and with agencies in Canada and Mexico. Officers of the state and provincial commissions enforce regulations to let hunters obtain a balanced "harvest" every autumn along each of the four principal flyways that migrants follow: the Atlantic coast, the Mississippi Valley, the central flyway (just east of the Rocky Mountains), and the Pacific coast. International cooperation is excellent and essential, for these feathered travelers make extensive use of breeding grounds in wetlands far north, touch down to feed (or to spend the winter) on national wildlife refuges in the United States, or progress onward into Latin America. The American program, which offers some benefits to nongame and endangered species too, is paid for by federal excise taxes and "duck stamps," plus state licenses, permits, and tags. In 1974, these fees cost resident sportsmen about $8 each, and nonresidents somewhat more.

The pheasant we see reached the New World first as a personal gift. They came in 1882 from Owen Denny, then U.S. consul in Hong Kong, for release on his brother's farm in the Willamette Valley of Oregon. Until then, no pheasant of any kind lived free in the Americas. Nine years later, the state of Oregon offered the country's first open season on cock pheasant. Some fifty thousand of them were shot in 1891 without impairing the breeding stock.

Other states imported the same kind of Chinese ring-necked birds. Soon they were reproducing in the wild and also spreading free from hatcheries set up from coast to coast. In the northern grain states and central Canada, the pheasant found so much food and shelter all year that the population rose to one bird per acre. South Dakota, Iowa, and Minnesota became the "ring-neck states." In a single year, hunters shot as many as sixteen million pheasant. This total exceeds that of all the wild ducks combined and afforded an amount of meat said to equal that from fifty thousand beef cattle.

By the late 1940s, South Dakota produced about 30 million harvestable cock birds without having to raise any in pens; the original 4,000 that were introduced started off this explosive increase. By 1963 the total had fallen to 10 million, yet they attracted no fewer than 68,901 nonresident hunters. These people spent $1,750,000 on licenses to shoot pheasant and an additional $11 million in the state on the amenities of their sport. Businesses flourished. But the bird populations continued to shrink. The total fell below 5 million in 1964.

Had pheasant been harvested excessively? No one believed this, for a cocks-only law left plenty of male birds to keep the protected females productive. Was DDT or some other pesticide causing nest failures, perhaps by excessive thinning of eggshells? The thinning was real, but not excessive. Were foxes and other predators to blame? But New York State had already tested the impact of foxes by eliminating fully 85 percent of these animals in one area and none in another of equal size just across a long narrow lake.

Pheasant populations were unaffected, for they reproduced equally well on both sides.

Richard C. Nomsen of the Iowa Conservation Commission believes he has the answer in the Middle West. He sees
a change in farming practices as reducing the habitat suitable for pheasant. In many regions, oats are the only field
crop left standing undisturbed until after hen pheasant
have hatched their broods among the grain plants and escaped. Now fewer oats are raised than in the 1950s. Hay
fields too are cropped earlier and more closely than previously, destroying uncounted nests of pheasant. Nor can
the pheasant make much use of marshy land, where they
used to hide and nest and feed. It has been drained or
filled, converted for field crops. (In Wisconsin, a third of
the wetlands disappeared in this way.)

Fall plowing, which is now practiced widely for greater
efficiency, buries the scattered grain upon which the pheasant formerly fed all winter. Even in Oregon's Willamette
Valley, where the ring-necks found their first success, their
population is down. It dropped as though on one end of a
seesaw, at about the same pace as the rise in use of clean
agriculture and weed-killing sprays, both of which oblit-
erate a pheasant's opportunities in life.

In New Hampshire and some other parts of New England, the fortunes of pheasant and of the native cottontail
rabbits have gone down together. These areas, which once
were heavily forested, became more hospitable to small
game as the trees were felled and farms took their place.
For a while, almost 50 percent of the New England scene
was used for agriculture. Now the farms are abandoned
and grown over with timber. Some hunters complain that
the cottontail has become an endangered species and the
pheasant a bird of the past.

The varying fortunes of hunters who search for trophies
and targets cause exultation in good years and frustration
in bad ones. Yet those who seek deer, upland game birds,
or even rabbits can scarcely expect the satisfactions of the

waterfowler. They benefit from no regular migrants arriving from the wilderness and rarely contribute to the income of the person upon whose land the hunt occurs. That land could be managed to yield game rather than developed to produce a cash crop, with wildlife finding habitat as best it can. But the biasing of the habitat for wildlife is unlikely unless the owner operates it as a hunting preserve, open only to those who invest a basic entrance fee and an extra charge for each animal they take away. By contrast, waterfowlers support the maintenance of wetlands that sustain millions of ducks and geese.

Tradition in North America surrounds waterfowlers with a most peculiar heritage. It was formalized in 1915 by the Migratory Bird Treaty between the United States and Canada, established for the British colony by a royal decree from King George V. Beneath the legal language lies a clear determination to manage the living resource in such a way that no migratory insectivorous bird, no game bird or nongame bird, would ever be hunted to extinction. The program has succeeded for more than sixty years and received high praise as a model for international cooperation. Its faults have become evident, but not a way to remedy them.

The Canadians, in particular, would like to see a revision of the treaty terms. No longer are aboriginal hunters—Eskimos and Indians—so few and ill-equipped that the treaty should allow them unlimited access to waterfowl at any season by any method. Currently some 1,500 of these privileged people are taking annually about 130,000 geese, in contrast to fewer than 37,000 geese shot by the 450,000 Canadians who buy licenses. The aboriginals now demand a guarantee of 85 geese each in every year and the right to insure this harvest by use of automatic weapons and rented helicopters as well as traditional nets, traps, and firearms. Should a court rule in favor of this development, which the framers of the treaty could not foresee? Such a decision might well lead to abandonment of management programs aimed at sustaining waterfowl populations, since these are paid for by license fees. Canada's extensive wetlands could profitably be converted into grainfields, ending most of the supply of geese and ducks that now wing southward over hunters in the United States.

The treaty of 1915 makes no provision for either legitimate complaints from one country to the other over management details that affect migratory birds or for revision to accommodate change. Officials of the U.S. Fish and Wildlife Service and of various divisions of the U.S. Department of Agriculture have no right even to make suggestions. They recognize at the moment that developments around James Bay aimed at producing electric power will soon reduce the number of Canada geese that migrate to winter quarters over the United States. Canada geese will not be endangered, even if their populations decline dramatically.

Canadian officials in the interlocking ministries of Environment, Fisheries, Agriculture, Science and Technology, Indian Affairs, and Northern Development cannot object through formal channels when activities in the United States affect migratory birds with breeding grounds in Canada. No support was given an informal suggestion that the famous whooping cranes, hatched in Wood Buffalo National Park near Great Slave Lake in Canada, would benefit in Texas during the winter by a simple change in policy. The Texas Fish and Game Commission declined to prohibit shellfishermen from working the waters between the vital Aransas National Wildlife Refuge and Matagorda Island, even if the cranes were disturbed. Nor would this commission and its counterpart in California consult Canadians regarding the harvest of white-fronted geese, although 90 percent of these handsome birds come from the Yukon and the rest from the Canadian Northwest Territories and Alaska. The white-fronts leave their far northern breeding grounds and come under the jurisdiction of a succession of states long before hunting season begins in Can-

The United States provides most of the food and refuges for migratory Canada geese from autumn until spring and Canada, the preponderance of food and breeding territory from spring until autumn. The hunting rights of licensed waterfowlers in both countries and the claims of Indians and Eskimos now seem to merit a revision of the amicable treaty provisions between the two countries without indicating what changes would be fair and practical in management. (Photo of adult goose by Rex G. Schmidt and of nest with goslings by Ray C. Erickson, U.S. Fish and Wildlife Service)

ada. Canadians see no way under the treaty to benefit from the birds that their wetlands produce. A fairer deal might be worked out, but only if the treaty were changed.

Many knowledgeable people in countries to which Canada might sell or give grain would like to see a new treaty

regarding migratory waterfowl that cross national boundaries. Protected ducks and geese now devour between three
million and five million bushels of wheat in Canada close to
the breeding grounds and along the broad flyways to the
south. Malnourished people could use that grain. Canadian
farmers could profit by harvesting it. These sacrifices are
made almost exclusively for the benefit of waterfowl
hunters in the United States. Should part of the income
from their license fees be used to reimburse Canada?
Would Canadian pride allow the acceptance of such a subsidy? Hungry people in the Third World could scarcely
approve of a program that overlooked so completely the
human needs of the poor.

Solving these problems of land use will be challenging.
Since 1931, Canada has had treaty-making powers, which
replaced the royal prerogative. But use of those powers
requires unanimous consent of all nine provinces to any
fresh agreement. Under these conditions, no quick change
seems likely. The United States is equally reluctant to replace the 1915 treaty, because the treaty terms have become the basis on which later accords were formulated. If
the document protecting migratory birds across the Canadian border were rewritten, new negotiations and agree-
ments would be necessary with Mexico, Japan, and the Soviet Union. As Canadian staff specialist Graham Cooch,
deeply concerned with migratory bird ecology, remarked to
us recently, "What a can of worms any new treaty arrange-
ments would open! So long as we can half-agree informally,
the welfare of all parties seems better served."

As we think about it, we wonder which pattern of land
use today represents the least digression from the old un-
steady balance of nature. A wide diversity of plants and
animals has little economic appeal. In any monoculture, the
cropland is devoted so far as feasible to producing a single
product, a commodity that enters the stream of commerce
and serves the needs (real or imagined) of mankind. In a
wildlife refuge or a game preserve, the native or foreign

animals are cultivated in a relatively natural environment to meet the demands of affluent hunters. Either enterprise collapses in a few years if untended. In neither is consideration given to continuation of complex ecological forces. Yet it is these that hone the genetic endowment of living things, making them what they are and influencing what they can be a millennium from now.

Perhaps there is no longer any possibility of sustaining the basis of old-style natural selection in the evolutionary processes of native plants and animals over most of the earth. The habitat has changed too much during recent decades, if not during the few millennia of mankind's advance to world dominion. Some sanctuaries remain. A few have been improvised. We feel surprise whenever the substitute is enough like the natural environment to enable the favored animals, at least, to reproduce there.

We think of "Père David's deer," the mi-lu, which has been extinct in the wild for no one knows how long. Its name commemorates the Abbé Armand David, a nineteenth-century French missionary who heard of the existence of these then-unknown mammals. A herd of them had shelter in the Imperial Hunting Park, in the confines of the royal palace in Peking. David bribed the guards to get him two skins, which he sent to Paris along with a written account detailing everything he could learn about the deer themselves.

Science might never have known more about these animals, had not Herbrand Arthur Russell in 1900 used his prestige as the duke of Bedford and the president of the Zoological Society of London to acquire from the herd in Peking about twenty healthy specimens. Russell made the animals at home on the grounds of his family estate, Woburn Abbey, and was delighted to see a rapid increase in their numbers. He shared his breeding stock with major zoos around the world but was saddened to learn in 1920 that the last of these deer in Peking had died "of old age." Russell's son, the present duke of Bedford, not only opened the grounds of Woburn Abbey to visitors, letting them watch the herd of Père David's deer; he also resupplied Chinese zoos with several pair in 1960, all descended from the original twenty animals.

We think too of the gentle little Hawaiian goose, the néné. Its survival on the volcanic slopes in national parks of that island state seemed so unlikely that a few of the pitiful remnant were caught and shipped to England. There Sir Peter Scott had offered the birds sanctuary and expert supervision on the grounds of his Wildfowl Trust, in Slimbridge, Gloucestershire. Imaginative care, including a special whole-wheat bread as a food supplement, soon had the birds breeding.

Two hundred Hawaiian geese had already been shipped off to Hawaii and to zoos all over the world before we met the thriving flock at Slimbridge. We cherish a photograph of one little goose daintily accepting a piece of bread from an outstretched hand, while another néné is sneaking a few good bites out of the rest of the loaf, held "out of sight" behind the donor's back. What would have happened to the néné, Hawaii's state bird, had the facilities of the Wildfowl Trust not been used to rescue it in its decline?

Other animals on the list of endangered and rare species are now breeding on remote reserves but are extinct in their homelands because of excessive hunting. Some, such as the white-eared pheasant of China and the Arabian oryx, are known presently from captive populations, which are breeding well. Others, such as the Asiatic lion, have lost their last native habitat to mankind and domesticated livestock. For all of these there is hope that healthy survivors will be available for reintroduction on their native soil when people who live there decide that they can share the world with wildlife.

The feasibility of reintroductions is supported by experiences in many lands. Failures on foreign soil give no measure of the probabilities on home territory. The European capercaillie, for example, has been freed in hundreds of

The gentle little Hawaiian goose (néné), once endangered, has been saved by special care near Slimbridge, England, on the grounds of the Wildfowl Trust, directed by Sir Peter Scott. A healthy breeding stock is being augmented at frequent intervals on the volcanic slopes of Hawaii where these birds are native. (Photo of two birds at Slimbridge by Lorus J. and Margery Milne, of three birds on the island of Hawaii by officers of the U.S. Fish and Wildlife Service)

exotic woodlands that seemed suitable. This grouselike bird is so huge—second only to the wild turkeys of the New World in size—that hunters on other continents have paid large sums to introduce a breeding stock. Rarely have the capercaillies survived a single season. Yet in Scotland, these birds made a quick comeback when Scandinavian capercaillies were released almost a century after the last of this magnificent species had been hunted to extinction in the British Isles.

The American turkey and the bobwhite quail have been tried in California, and the California quail in eastern states. Neither has succeeded so far from the areas where these birds lived a century or two ago. Yet turkeys thrive almost anywhere in their native range so long as hunting pressure is kept low. "Low" means the equivalent of 0.17 hungry Indian to the square mile and hunters with no firearms of any kind.

The time seems ripe for changes in land use, toward perpetuating life rather than causing it to disappear, toward enriching its variety and ecological interactions rather than simplifying them. Now fascinating tests are under way toward reinstatement of animals that disappeared, hunted out as targets and trophies. The French government is studying the possibilities within the new Parc du Vercors, southwest of Grenoble. It seems to have habitat for the European brown bear (close relative of the American grizzly), the lynx, the ibex, the chamois, the marmot, the golden eagle, and the magnificent bearded vulture.

Each year, each month, each week it seems, the news tells of a fresh success with some form of life that human actions have decimated. The Hawaiian geese are nesting again on their native island. The musk-ox has been reestablished in Siberia, after centuries of being extinct in the Old World. South Africans have saved the bontebok (a handsome antelope with an almost purple coat, white face and rump, and stockinged feet) and the white-tailed gnu. The

Wild turkeys attain trophy size in parts of North America where formerly they were numerous, but only if these birds live in suitable habitat and receive a degree of protection that many hunters regard as unreasonable. Populations of wild turkeys decline and disappear if the human "harvest" exceeds the rate that was customary before any Indian had firearms and while the population of these hunters remained less than two per square mile. (Photo by Luther C. Goldman, U.S. Fish and Wildlife Service)

Sinai leopard has been photographed alive and well where it had long been given up as lost.

It is still too early to speak of "triumphs," for all conservation measures serve first as delaying actions. They postpone the day of disaster. At best they confer upon coming generations of mankind the responsibility for deciding whether some form of life or piece of habitat merits preservation. Once the species is extinct or the habitat has been destroyed, this option disappears. Conservation keeps options open, without promoting any ethic or value system that might give guidance the next time around.

We can think along with the gracefully outspoken naturalist Aldo Leopold, who pioneered wildlife management. He came to doubt the strength of any conservation system based wholly on economic motives, but he lacked the proof he needed to link most nonhuman life to the survival of mankind. Probably no more than 5 percent of the twenty-two thousand different higher plants and animals native to

The musk-ox is well adapted in body and behavior for life in the Far North, remote from people. Yet the animal can tolerate the climate in New England and adjust to human management as a semidomesticated source of meat and fine hair. Musk-oxen raised in North America have now been shared as breeding stock with the people of Eurasia, where the animals were exterminated long ago. (Photo courtesy U.S. Fish and Wildlife Service)

Leopold's state of Wisconsin could be "sold, fed, eaten, or otherwise put to economic use. Yet these creatures are members of the biotic community, and if (as I believe) its stability depends on its integrity, they are entitled to continuance."

Recent discoveries have revealed the dependence of a stable, outbreak-free community of life upon its own complexity, upon the lesser plants and animals rather than the target species or trees of commercial worth. The health of the outdoor world can be measured best by noticing inconspicuous signals. The purity of flowing water is revealed by its mayfly populations, and the quality of unpolluted air by the kinds of slow-growing lichens that cling to rocks and trees. The diversity of butterflies in a biotic community shows more reliably than most other indicators its suitability for human and nonhuman life. Obviously or otherwise, for better or worse, we each share in a local living world. Any deterioration shows among the members of no known economic worth as an early-warning system apprising us of our own future.

Our friend and mentor, the critic and writer Joseph Wood Krutch, recognized the futility of piecemeal effort and concluded that conservation is not enough. It does not safeguard the living needs of future members of our species, which will equal or exceed those we feel. These needs will not be less when the world is called upon to support 8 billion people. Coming generations should be left every option possible and every constructive idea about how to manage on a planet with less of everything than we ourselves inherited.

6

The Clock Moves
Only Onward

IN MOST PARTS of the world, it seems, todays become tomorrows faster than they used to. Yet history never quite repeats itself. Essentially the same sequence of events may transpire first in one place and then in another—but not twice in the identical area. Always that area has changed. It affords a new combination of opportunities. Its tomorrow can never be yesterday.

Each renunciation of a familiar way for a new one begins a perilous adventure into the unknown. Virtually no chance is ever left to say, "Enough! I liked the old way better, and I want to return to it." Rarely can an Eskimo in the Arctic turn back after setting aside the successful dependence on seals, white whales, and caribou to work for cash. More money is always needed at the sales counters, for a poorer diet, less suitable housing, and an unfamiliar life that combines initial excitement with many frustrations. Performance of unskilled labor, accumulation of a bank account, an impersonal role toward gross national product take the place of skilled, unpaid, self-supporting activities with minor impact upon the environment.

The seals, white whales, and caribou may gain from human neglect. Certainly the caribou near the new pipeline

Caribou thrive in the Far North if they can travel between winter shelter among the spruce and fir trees of the taiga where they find an abundance of lichens, and summer calving and grazing territory on the arctic tundra. Domesticated caribou (reindeer) in northern Eurasia must be kept on such a schedule by herders. Habitat destruction and either too little or too much "harvesting" of caribou tend to destroy populations of these arctic deer. (Photo by Robert D. Jones, Jr., U.S. Fish and Wildlife Service)

in Alaska prove unexpectedly tolerant of jet aircraft landing and taking off, of road machinery passing close by, of people deprived totally of guns and alcohol. The animals walk the pipeline road, undaunted by this ribbon of tamped gravel three feet thick, built to hold traffic above unwarmed permafrost. This too may change irreconcilably if and when access to this remote area is opened to anyone with a wish to go there.

Patterns of civilized behavior in the nuclear age press upon primitive peoples in many places. Old ways of life well suited to the natural resources of the region in which these people live are fading out. The marvelous products of technology seduce primitive peoples into abandoning their traditional culture and trying to adopt a substitute. Or the strange law of eminent domain may be applied by civilized colonists to determine what is to be done with undeveloped lands and their native populations. A few months is enough to restructure a whole community, upsetting its old balance.

Even in a smoothly regulated society, the impossibility of turning back the clock becomes apparent. Scientists in New Zealand first challenged us to consider the effects of this "progress." They specifically disbelieve the statements they read about bared land in the Northern Hemisphere: that it will be colonized by a succession of plants in a predictable sequence wherever a farm is cleared and then abandoned. Experience in New Zealand completely refutes the idea that what was once woodland will revert to its original state in fifty or a hundred years, as though nothing had happened. How could stone walls and cellar holes vanish under a mature forest in New England if farms, once cleared, were allowed to grow up again?

"Do you mean that the community of New Zealand plants and animals reaches no steady state," we asked, "or just that the ecological succession takes longer—longer than anyone has been able to observe?" We are used to progressive changes until a self-sustaining climax community is reached, in which the plants and animals are those whose young can take the place of old members of the same species.

"We see no 'steady state' anywhere. Everything keeps changing. No community of life in New Zealand consists of species that replace themselves without major additions or losses. No climax community exists."

Later we had an opportunity to see one of the remaining forests and inquired further about the great kauri "pines" that tower to 180 feet above low, handsome members of the laurel family. "What colonizes the exposed area of forest floor after the valuable kauris are cut?"

"Manuka," came the quick response. *Manuka* is the Polynesian name for the teatree, just one among thousands of Maori words the New Zealanders adopted.

"That's a shrub?"

"Ah, yes."

"Can a kauri seedling grow in the shade of a manuka bush?"

Yes again. But who will wait for a kauri to reach commercial size, in ten to twenty centuries? True pines and other quick-growing trees from the Northern Hemisphere are planted to replace the giants, despite their special resin and quality of wood. We marveled to see tracts of Monterey pine in New Zealand, above a dense understory of tree ferns. In California, Monterey pine is outranked in value by most other conifers.

We looked at the soil under the new stands of pine. The litter of needles and broken branches was familiar enough. But underneath lay a peculiar soil, one we had never encountered previously. Its source was fallen foliage of laurel trees, and shed bark and flat "needles" of kauris that had been felled and hauled away.

Any soil is dynamic. Litter nourishes it. Fungi and bacteria decompose it, with the help of many inconspicuous animals. Eventually it becomes unidentifiable organic matter (which retains moisture) and mineral substances (which roots absorb or groundwater leaches away). Beneath the new pines, the soil will change gradually, becoming more acid and less able to renew its fertility. What then will grow there? Will New Zealand insects and birds, which once found food and hiding places within kauri forests, live in some new locations on different resources? Or must this lovely land be poorer for their loss? A century hence, when it is too late to dream of turning back the clock, New Zealanders may think their pine products poor recompense for a rich birthright.

A somewhat similar situation exists in South Africa, where tree farms of Monterey pine now replace quite different native trees, such as those known as stinkwoods. One is a member of the laurel family, a second closely related to our hackberry. Both grow very slowly, producing a heavy wood with extremely fine grain, used and admired by many South Africans for massive cabinetry in their homes and public buildings. People can alter their taste in interior decorations. Can the other plants and animals, which lived in the wild community with huge stinkwood trees, now fit into other habitats? Among these displaced creatures are several of the world's nineteen kinds of touracos. Only Africa is home to these handsome, noisy, elusive birds of the tall trees. It is there that they court, mate, nest, and learn to fly, although they also descend to lower vegetation to feed—so often to the wild plantains of the banana family that the birds are known also as plantain eaters. Several are now on the list of endangered species, threatened with extinction.

In the Northern Hemisphere, native trees that possess prime value are used for reforestation by big corporations that harvest timber. A mixed stand of spruce and fir on western mountains is often clear cut and then planted with a selected stock of Douglas fir. The stock is chosen for resistance to insects and disease and the ability to grow quickly and straight among others of its own kind and age. The corporations rest their advertising claims upon such programs, which now produce new trees twice as fast as they cut old ones. Rarely is mention made that the quick-grown wood is softer and less valuable than that in old-style slow-growing trees of the same kind or that the new man-made forests are of a single species with a limited appeal to native animals or people interested in outdoor nature. The even-age stand is to be cut as soon as it matures, because then its growth rate lessens. A new crop will be planted. This is a production forest, not one with a multiplicity of values. True, it will produce wood and oxygen, retain precipitation, and hold soil. Yet the soil will progressively reflect the fact that it nourishes and receives contributions from trees of a single kind. Other kinds of vegetation, like the wildlife they could support, have no place in such a forest. Its simplicity puts the ecology out of joint.

These changes alter the environment less drastically on any well-watered, temperate part of a major continent than on an isolated island. Yet the immensity of the change is often concealed from the mobile traveler. Events on the ground cannot be interpreted meaningfully from passenger aircraft because airplanes fly too high and fast. Automobilists tend to follow paved roads, especially when driving through forest. Recently we varied this pattern by turning off the highway that leads to Olympic National Park in Washington State. Inviting byways led through the tall trees on both sides of the road. In fact, their frequency puzzled us, for they were only a few hundred yards apart. We found the answer as soon as we progressed down one of these byways less than an eighth of a mile. Only a thin corridor of forest remained along the highway, and beyond it lay a vast area with only jutting stumps, scattered slash, and a confusion of upright leafy stems growing from the surviving roots of felled trees. The impressive vista from the highway was a front, as false as a movie set. How much wildlife and how many native plants could the false front shelter?

A roadside that is not mowed or poisoned selectively with weed killer often becomes inviting to animals with short legs, because they find low vegetation there, including a considerable assortment of vigorous weeds. Often the roadside is the principal remainder from a former continuum of native grassland. Yet it is hazardous, most obviously because the creatures are likely to be crushed if they wander onto the pavement. Their carcasses attract scavengers; these too may be struck by vehicles traveling faster than most animals can. Most passenger cars add inconspicuously to the unintended carnage because their exhausts distribute a lead dust from the tetraethyl lead in gasoline. Roadside rodents and rabbits eat the vegetation that absorbs the lead; lead poisoning slows their reactions. Scavengers that feast on the carcasses concentrate the lead until it affects them too.

We often think of this when we notice on the automobile ahead of us a bumper sticker reading Warning: I Brake For Animals! Does the driver show consistency by choosing a lead-free gasoline? And what of his brake linings, which add to roadside dust a measurable amount of asbestos fibers each time the brake pedal is pressed? Environmentalists who measure these contaminations are first astonished and then dismayed by the continued accumulation of real hazards along our highways.

Just occasionally, someone recognizes an opportunity to deflect the adverse influences of mankind at comparatively small cost and without inconveniencing as many people as could benefit by the change. Both aspects of improvement hold great importance. In 1973, Daniel F. Jackson, environmental technologist of Florida International University, introduced his students to such a situation close to Miami. It became possible because the city of Miami had installed a low tidal dam to prevent the inland intrusion of salt water up the canal known as the Miami River. During rainy seasons, this channel serves as one aqueduct among many that now drain Lake Okeechobee. The whole system, designed and managed by the Army Corps of Engineers, keeps the water level almost constant in agricultural land surrounding the big lake. In drier months, the Miami River shows little flow. Yet its water remains surprisingly clean, suitable for fishes and air-breathing wildlife, until it reaches the northern edge of Dade County.

Within Dade County, Jackson's students traced a progressive rise in surface oil, asbestos fibers, lead salts, and coliform bacteria (which indicate sewage contamination) as the water shifted toward the tidal dam in Miami. Correspondingly the habitat deteriorated for fish and wildlife. People regarded the channel as little more than an open sewer and dumped refuse in it. Maintenance crews came periodically with heavy equipment to remove new accumulations of water hyacinths and submerged vegetation, because these clog the river and reduce its effectiveness in carrying runoff. About a dozen houseboats were permanently tethered to the bank. Otherwise the only recreational use the river served was as a straight raceway for people in boats with outboard motors.

The asbestos fibers and lead were easiest to trace. They increased after each rain as the water flushed the pavement of Okeechobee Road, which parallels the river through most of Dade County. On the side next to the channel, the gutter collected the water and pollutants and discharged them through multiple drains directly into the river. Investigation revealed that on the opposite side of the same road, the gutter sent similar washings into a big storm sewer, which eventually reached a waste-water treatment plant in Miami. When this difference and its consequences were drawn to the attention of the mayors and other officials of the communities along Okeechobee Road, instructions were given to the public works departments to modify the system whenever any work is needed on gutter drains so that all of them discharge into the sewer. Gradually and at minimum expense, this source of river pollution disappeared.

The coliform bacteria count was properly the concern of public health officers in the several communities, for it formed the basis of a law forbidding people to swim in the Miami River. Yet no sewers were known to empty or even leak into the channel. Students discovered that the coliform bacteria count peaked on Mondays, sagged through Friday noon, and then rose again; it crested close to each tethered houseboat. Houseboats served mostly for weekend parties, and sewage came from each one, untreated, into the river. When the mayors and public health officials were furnished with this information, a dilemma arose. How could houseboat-owners be required to install holding tanks and to empty them at waste-disposal sites if no sites existed?

How could construction of sites be justified if no one was required to use them? Yet more voters would favor a correction of this abuse than would be hurt by ending it. The situation was resolved by notifying all houseboat-owners that their licenses to maintain their craft would not be renewed unless they installed holding tanks and regularly emptied them at disposal sites; two such sites would be built in time for projected use. All except two of the owners reacted by removing their houseboats before the date when their licenses would require renewal. The two exceptions chose to install pumps and sewer connections to shore facilities on their own property, rather than put in holding tanks and have to move the tethered houseboats to waste-disposal sites at frequent intervals. Coliform bacteria practically disappeared from the Miami River at no cost to the communities and their taxpayers, because the disposal sites were never built!

The surface oil continued to interfere with small animals and water birds until a new ruling was agreed upon and enforced by all of the communities along the channel. Only hand-propelled boats might be operated on the waterway. Almost abruptly pollution vanished. Ducks of several species and wading birds (such as herons and egrets) began to frequent the river. Fish came downstream, and so did kingfishers. The channel maintenance crew took out time from their continuing battle with water hyacinth to show us where two different mother ducks were incubating eggs in riverside sites.

Citizens organized the First Annual Miami River Regatta in 1974. Canoes raced up the channel and back again. Seminole Indians accepted the invitation to erect palm-thatched shelters along the river bank in Miami Springs, close to the finish line. They added a colorful, festive character to the celebration and did a thriving business in handicrafts with visitors to the regatta.

The city councils along the channel discovered an amazing number of undeveloped narrow lots that could be inexpensively made into picnic sites, just where people could watch the birds and canoeists. Fourteen miles of bicycle paths along the river were laid out. Suddenly the communities became protective of their unexpected recreational resource.

Professor Jackson orchestrated these studies and encouraged repeated contacts between his students and community officials. At the end of each term, he scheduled an evening presentation by the students on specific findings and the actions taken. Each report had to be brief, well illustrated, and a real contribution to the studies. Not only the officials of the local communities and the interested public were invited, but also representatives of state government. Each delegate who came (and each office that sent no one to the presentation) received a spiral-bound volume of these lively reports, every section credited to the students who did the actual work. Each student received a copy and saw the advantage in taking it along if applying for employment in any of the community offices or state agencies that had been kept informed of the Miami River studies.

Mutual respect and enthusiasm stay at a high level. New ideas come forward and get explored for feasibility with modest funds. The tidal dam on the Miami River, for example, remains an obstacle for fish and sea cows. Ordinarily they swim on their own schedules into the salt water as soon as winter weather chills the river and swim back again in spring. Tarpon and other fishes might flip themselves up or down around the main barrier if a fish ladder were built. But what of sea cows seven to twelve feet long, weighing as much as fifteen hundred pounds? One of Jackson's ingenious students devised a sea-cow escalator that could be built for less than $5,000. It would gently lift these ponderous animals up and over the tidal dam one at a time. The sea cow would need only to indicate interest in going in either direction by nudging a trigger at the end of the motorized rubber apron. Once in the modernized Miami River, the

sea cow would no longer be in danger from the sharp
blades of propellers on motor boats, which can easily cut a
succession of deep wounds into the animal's back. People
along the waterway would be fascinated to see these legen-
dary mermaids munching on water hyacinth, keeping the
channel open gratis, a substitute for expensive maintenance.

One major need was evident in these studies: a laboratory with electricity and running water close to the river,
where students could work on their samples. The Jaycees—
the Junior Chamber of Commerce—filled the need by offering an unused fourth of their building only a block from
Okeechobee Road. Then these businessmen helped Jackson raise a few thousand dollars to equip the laboratory ad-

The slow-moving sea cow (manatee) comes to the surface by day
only for an occasional breath of air. By night these animals feed
on submerged vegetation or reach out of the water for overhanging branches and grasses along the shore. Some people dream of
making the world safe again for sea cows, to gain the services of
the live animals for aquatic weed control and their meat as a reliable food for mankind. (Photo courtesy of Miami Seaquarium,
Florida)

equately, since funds from the state university could not be spent on property that was not state-owned. Students provided most of the labor. At a record pace the Potamological Laboratory of Florida International University came into existence.

Everyone shares the excitement of seeing what can be done and what remains to be accomplished. Each gain uncovers other possibilities, all within the local area. Yet every change must be achieved in a new context, benefiting both people and their environment.

In many parts of the world, the causes of a deteriorated habitat are either so many or so inaccessible that correction is impossible. The only practical approach may be to improve the local situation so completely that it compensates for detrimental factors beyond reach. One of the most inspiring examples of this extra care at close hand is progressing in Bermuda. These islands lie about 570 miles east of Savanna, Georgia. This is farther from the tropics than coral usually grows, yet coral there perpetuated the site of a volcanic cone and let it accumulate hard layers of windblown sand. Soon after its discovery in the early sixteenth century, Bermuda acquired an evil reputation among sea captains, mainly because of a lack of flowing springs of fresh water. Yet a few ships stopped, and European hogs were freed ashore to multiply as free meat for any later visitors. No settlers, however, arrived to alter the islands further until the next century.

The first people actually to live on Bermuda came by accident. Their ship, the *Sea Venture,* was bound for Virginia. A terrible storm battered the vessel for four days, and Captain George Somers barely managed to wedge his leaking craft between two rocks near an island shore. By nightfall, all 150 men, women, and children were huddling together on the beach, close to a dense forest of endemic cedar trees, terrified both by the stories they had heard of these "isles of ill omen" and by wild clamor that seemed to prove the presence of malevolent forces.

The morning sun revealed a less dreadful reality. Several diaries record that "all the Faries of the rocks were but flocks of birds, and all the Divils that haunted the woods were but heards of swine. . . . Fowle there is in great store," especially seabirds (petrels) that produced terrifying calls at night. The birds, like their eggs and young, proved edible and so unwary that dozens could be killed with a stick. "The Shoare and Bayes round about afford great store of fish. . . . Turkles [turtles] there be of a mighty bignesse. . . . The country yieldeth divers fruits. . . . It is, in truth, the richest, healthfullest and pleasing land."

Still, it was not the people's destination. And when rescue came, they continued their journey to the mainland. Their accounts, however, led King James I to charter the Bermuda Company in 1616 and to send a shipload of colonists and supplies there as an outpost of Virginia. The Britishers brought pet dogs, cats, horses, cattle, chickens, and seeds of agricultural plants to make the island life like that of England. Rats, mice, and houseflies, which traveled along as stowaways, quickly settled in.

For a few years, the turtles came in season to lay their eggs. The seabirds arrived to nest and raise their young, but the colonists and the rats killed so many of them that soon no more could be found. The wild pigs were slaughtered too. House cats reduced the number of white-eyed vireos, which nest in shrubs close to the ground, and almost exterminated the little skinks—the only reptile native to the islands. Vireos and skinks were the principal predators on houseflies, and with few of the predators remaining, the insect populations burgeoned. By 1875 the Bermudians decided to remedy two complaints at once: too many houseflies and too few familiar birds. House sparrows were introduced from England, as though these Old World weaver finches (which are chiefly seed-eaters) would destroy more houseflies than the remaining vireos, the native bluebirds, or the warblers of thirty-three kinds that come through Bermuda each autumn, hungry from a long migratory flight over open water. The house sparrows soon caused a decrease in the number of bluebirds, by occupying every nest hole big enough to squeeze through. Fortunately, bluebirds are slightly slimmer and found some nest sites that sparrows could not enter.

Other foreign eaters of insects were introduced. They included in 1880 the sweet-voiced whistling frog of Martinique, via Jamaica; this amphibian is active at night while houseflies sleep. The giant "marine" toad of northern South America, added in 1885, prefers large insects such as cockroaches. It eats also small birds, small fishes, even cat food. Any cat or dog that picks up this amphibian in its mouth risks getting a fatal dose of poison from the skin. (Australians are currently trying to exterminate some of the same kind of giant toad that escaped after a school teacher in Darwin got some for his classes.) Bermudians freed in 1905 some six-inch *Anolis* lizards from Jamaica; these actually can catch and will eat a housefly. The imports modified the living community in ways that made it vulnerable to a series of subsequent disasters.

The principal culprits arrived in 1944, undetected on some decorative shrubs from the United States. Both were scale insects, one of them a near relative of the dreaded oyster-shell scale. Actually, it is a pest chiefly of oriental umbrella-fir. It seemed the more dangerous kind when it was discovered in enormous numbers on branches of the local Bermuda cedar trees, killing them. Accompanying this pest was another scale that few scientists had met before. On Bermuda cedar it reproduced rapidly. Probably it caused the total disappearance of its competitor by 1948. By then, most of Bermuda's two million cedars were dead too. As these evergreen trees lost their needles, the privacy of the close-set houses disappeared. No longer did the trees, some of them fifty feet tall, act as an efficient windbreak. Gales from the surrounding ocean whipped unobstructed across the islands. They shredded the leaves of banana plants, drying them and ending the production of

Giant toads from tropical America have been introduced in southern Florida, northern Australia, and many islands. Dogs that mouth the toads may die of poisoning from the skin. The toads themselves prove far more versatile in accepting unusual foods than anyone expected, and they affect the welfare of many native animals that are small enough for them to eat. (Photo by Lorus J. and Margery Milne)

respectable crops of fruit. Now Bermudians had to import bananas—a new experience. No longer did their paradise sustain the illusion of being tropical, in keeping with the coral reefs offshore.

The Bermuda Department of Agriculture consistently rejected proposals to use poison sprays because the toxic mists could contaminate the large catchment areas where rainwater is collected for storage and later use. An appeal for help went to Canada. Soon a team of entomologists arrived. Why not release a few million harmless ladybird beetles? These small predators eat young scale insects at the only time they are unprotected: as they leave their mother's convex scale and creep to new locations, where they will suck plant juice while growing a scale of their own. By spring of 1945, nine technicians in Bermuda had thousands of ladybird beetles to release. They went free on living cedars all over the islands. But the Canadians had not taken into consideration the *Anolis* lizards, for Canada has none of these reptiles. The lizards on Bermuda promptly ate the beneficial ladybird beetles. In places where the small native skink had escaped the house cats, it too dined on the introduced black-spotted orange beetles. So did white-eyed vireos and bluebirds. Unchecked, the scale insects multiplied. Bermuda's cedars continued to die.

By 1949 all hope of control was abandoned. Bermuda's House of Assembly decided that money would be spent more wisely on removing the unsightly gray trunks of dead cedars than on trying to protect the remaining live ones. Funds were voted to clear dead cedars from government-owned land. Three years later an additional allotment went for removal of the tree specters within fifty feet of all 180 miles of Bermuda's roads. By 1955, when fully 90 percent of the cedars on the islands were dead, the swath was widened to one hundred feet on each side. Today, with only about 1 percent of the original cedar population still clinging to life, the mournful task is about complete. Horticulturalists wonder whether the surviving cedars are resistant to the scale insect and might pass on resistance to a new genetic strain of cedars. Meanwhile, foreign trees provide a windbreak: allspice from Jamaica, fiddlewood from the Lesser Antilles, Surinam cherry, Brazilian peppertree, and Australian "pine" (*Casuarina*) inherited the role.

A visitor might expect that Bermudians would recognize the vulnerability of their island plants and animals to introduced species and would forbid the addition of any more. Legislation, however, would not have protected native life forms from the European starlings. These managed to reach Bermuda on their own wings in 1954 from the United States, where they are regarded as a pest. The first of these aggressive birds in the New World were released in 1891 in New York City. In Bermuda the starlings began at once to compete for nest sites with house sparrows and bluebirds.

The conservation officer, David B. Wingate, launched a counterattack by promoting construction and maintenance

The eastern bluebird of North America nests in holes, those made either by woodpeckers or by builders of nest boxes. Introduced house sparrows and European starlings compete so vigorously for nest sites, where formerly the bluebirds were likely to encounter only native tree swallows, wrens, and chickadees, that bluebirds have become scarce. People in Missouri and New York, who claim the bluebird as a symbol of their states, see far fewer today than in former times. (Photo by Jack Dermid, U.S. Fish and Wildlife Service)

of birdhouses designed to admit bluebirds only. He knew, of course, that *Anolis* lizards from Jamaica had compounded a mistake. A second kind of these lizards was introduced from Barbados in 1953, and a third, from Antigua in 1956. But the conservation officer could get no legislative backing to prevent the introduction of kiskadee flycatchers from Trinidad, supposedly to eat up the lizards. These powerful birds, natives of forest edges from Argentina to southern Texas, not only attack darting lizards and flying insects but will catch small fishes in surface waters, fresh or salt. On Bermuda, kiskadees attack young bluebirds and kill and eat them. Now kiskadees outnumber bluebirds on the islands, and still no law forbids further additions to the harassed native fauna.

Wingate was responsible for discovering in 1951 that the native Bermuda petrel (called the "cahow" in imitation of its nocturnal cries) had not been utterly exterminated in 1621 by the early settlers and the rats. A few pairs are still breeding on the most remote islets of Bermuda, but the birds emerge from their nest burrows only in darkness. Even there the loss of the native cedars had an effect. No longer did a firm mesh of cedar roots resist erosive loss of

The kiskadee flycatcher of coastal Mexico, the West Indian islands, and southward proved an aggressive addition to the birds of Bermuda, where this predator on large insects was introduced in the hope that it would suppress the growing number of introduced lizards. The kiskadee preferred other foods, which affected Bermuda in ways that had not been predicted. (Photo by Luther C. Goldman, U.S. Fish and Wildlife Service)

soil between the coral rocks. For lack of a suitable turf below which nest cavities could be excavated, the petrels began to seek seclusion over the rim of the cliff, wherever they could find deep horizontal crevices in the rock. This move, however, brought the petrels into confrontation with a different bird of slightly larger size, the yellow-billed tropicbird, which Bermudians call the "longtail." Although the petrels arrive from their wanderings over the open sea to begin nesting a few weeks earlier than the tropicbirds, priority does not protect the petrel eggs or chicks. Returning tropicbirds seek out nest sites they have used before and eject any petrel they find inside, destroying eggs and chicks, sometimes killing the adult as well.

Wingate sought to bias the outcome in favor of the rare petrels. First he tried wedging a vertical diaphragm of plyboard in front of each occupied petrel nest in a cliffside location, providing only an oval hole big enough for a petrel to squeeze through. He reasoned that the tropicbird would not enter and thereby underestimated the territorial drive of these fish-eating birds. A newly mature tropicbird, hunting a place to nest, would bypass a diaphragm behind which a petrel engaged in family duties. But an older tropicbird, returning to a nest site it had used before, would force itself through any opening a petrel could use. The only way to provide a secure nest site for a petrel in the territory of the common tropicbird was to remove the latter, block the access opening completely, and install a perforated diaphragm the following year before petrels arrived to seek seclusion.

We accompanied Wingate on one of his frequent inspection tours of the outer islands to check on petrel nests. He showed us his preferred subterfuge on behalf of the rare cahow: an artificial nest burrow, constructed of wire screen and cement, in locations above the cliffs where formerly the cedar roots retained a thick soil and petrels burrowed. Each artificial burrow has a small entrance hole, a tunnel section a foot or more in length, and a nest cavity almost a foot across, capped by a removable cement cover. Gravel gives the nest a small measure of concealment. With the cap off, Wingate could see by the light of a flashlamp whether his petrel house was occupied in season. One was, and we managed to get a flash photograph of a fat chick inside.

The need for such elaborate efforts to save an endangered species from final extinction was recognized in the 1960s when the annual census of Bermuda petrels and tallies of their reproductive success showed a steady decline. Since various fish-eating birds, such as brown pelicans and bald eagles, were similarly decreasing in numbers because of accumulation of DDT and other persistent pesticides, tests for these poisonous substances were made on unhatched eggs and dead chicks of the petrels. Alarming concentrations proved to be present, although none of the materials are used on Bermuda. Clearly the parent petrels were absorbing the poisons from fishes taken on the open oceans, perhaps far away. Chicks got an overgenerous share of pesticides in the yolk of the egg and supplements from the partly digested fish the adults regurgitated as nourishment within the burrow. Bermudians could do nothing to protect their rare petrels from this worldwide hazard. The only local measure would be to give the endangered birds the fullest possible security on their home islands. Any surviving petrels might later have a chance to expand their population if the ocean environment responded to a lessened use of persistent pesticides in distant lands.

Wingate had another seabird to show us nesting on those outer islands. He looked under several piles of coral rock low on the shore before he found one; an Audubon's shearwater. This bird, which is slightly smaller than a Bermuda petrel but closely related, once numbered in the thousands at least and probably contributed a major share to the terrifying calls the shipwrecked people from the *Sea Venture* heard in the night. The shearwater took the name "pimleco" on Bermuda, before vanishing from this most distant outlier of its primarily tropical range. The bird is the "diablotin" ("little devil") in the West Indies, where it nests in shallow burrows and on the sides of precipitous cliffs. By day the parents shield their single white egg or dark chick under a black body and then noisily exchange places with a mate at night. Only the fact that the world has so many more Audubon's shearwaters than cahows conceals the impact of pesticides from far away on these seabirds as well.

Presently the petrels and shearwaters of the world, the fulmars, murres, small auks, and puffins of northern seas, and the albatrosses and penguins of the Southern Hemisphere are all diminishing in population. Neither pesticides nor oil spills are the new villains, although both continue to create difficulty. The current confrontation arises from outright competition for fishes as food. Human demand,

Nesting success of the bald eagle decreased in recent years until the U.S. national bird had to be listed as an endangered species. The reduction in numbers occurred in part because of the destruction of nest trees and of shallows where eagles could hunt for fish. Failures arose also because the concentration of pesticides discarded into coastal waters moved upward through the food chain into fishes, which are then impaired enough to become easy catches for eagles. Only in Alaska have these environmental changes remained minor. Shown here are an adult (photo by Luther C. Goldman), four downy nestlings beside partly eaten fish in the nest (photo by Frank Beals), and three growing juveniles in dark plumage (photo by V. B. Scheffer, all U.S. Fish and Wildlife Service).

despite higher prices, is no longer met by fishermen who
shuttle between fish piers in coastal cities and waters close
by. The industry cannot now equal its former harvests that
way. Yet it is called upon to supply a growing population. It
uses modern ships equipped with echo sounders. These
show the location of schools of fish where trawls might
catch them far below the surface.

Progress has required a different fleet of vessels,
equipped for pelagic operations. Fishes are now sought
along coasts that are scarcely inhabited by any people. Profitable quantities are coming from waters off the forbidding
shores of Labrador and Southwest Africa, where fish resources previously served seabirds almost alone. The birds
still come. They dive after fish between the fishing vessels
and get caught in the tremendous drift nets. An estimated
half-million birds each year drown while entangled beneath
the surface. Mortality from this new hazard probably exceeds that from all other causes combined. It is greater
than the production of young birds under the best conditions. Bird populations dwindle inexorably, with no direct
contact between hungry birds and fishermen.

Double jeopardy is the heritage of creatures that migrate

regularly between a breeding area and a feeding region they occupy off season. Their welfare can deteriorate at either end of the migration route, as well as in between. Even when the site and nature of the change can be identified and the survival of an endangered species is at stake, it may be difficult to redirect a program of development for human benefit to save the migrants. Thus, people who are clearing brushland in the Bahamas to make room for condominiums and high-rise apartments take little interest in the small insectivorous birds that have used this habitat for millennia. Bahamians can scarcely better their economic situation by catering to little birds, even if many of them are Kirtland's warblers, even if this is their only wintering area, even if the species is on the world list of species close to extinction.

The timing of this change in the Bahamas seems ironic, for it threatens to nullify a conservation program that held high hopes of success in northern Michigan, where Kirtland's warblers all go to nest. There a schedule of controlled burning is in its second decade, renewing the jackpine habitat these birds choose as breeding sites. A program for trapping parasitic cowbirds, which are not native to this part of North America, has diminished markedly this threat to the warblers' reproduction. There is even a spinoff that local people enjoy: a tremendous spread of highbush blueberries, heavy with fruit in season, all through the warblers' nesting grounds. We joined the people gathering fruit and saw that they filled their pails without disturbing the rare birds. This whole region will probably be allowed to grow up as forest land if the loss of winter habitat in the Bahamas obliterates the bird toward which Michigan extends such considerate care.

Far less specific changes in Latin America affect many kinds of migrants that travel for the winter to tropical forests that are being felled and to tropical wetlands that are being drained or filled. Instinct guides these birds to places where neither shelter nor food is now available. Failure in either direction soon obliterates the genetic heritage by progressive decrease in the migrant population. Nor do the political leaders or the land developers heed the warnings of scientists who can ignore national boundaries and recognize when a costly blunder is in the making. Most new attempts to convert well-watered lands in the Torrid Zone are doomed to fail as soon as the developers receive their pay. The soil is not forgiving. The highly adapted community of native plants and animals does not recover when the project is abandoned. They, like the migrants that have shared this complex tropical world for millions of years, can disappear in a few generations. With them goes the genetic heritage that is now seen by conservationists to be the greatest treasure of life on earth.

Conservation of the habitats in natural areas and of the genetic material they contain provided the focus of an international conference held during 1974 by the United Nations Educational, Scientific and Cultural Organization at

Period	Operator	Objective	Target	Time Scale
to 8000 B.C.	hunter-gatherer	the next meal	wildlife	1 day
to A.D. 1850	peasant farmer	the next crop	domesticated plants	1 year
from 1850	plant breeder	the next variety	same	10 years
from 1900	crop evolutionist	to broaden the genetic base	same	100 years
today	genetical conservationist	dynamic conservation	wildlife	10,000 years
	politician	current public interest	same	next election

Morges, Switzerland. Many topics received careful consideration. But Sir Otto Frankel of the British delegation sought to provide a historical perspective for the changes in human impact. He summarized his analysis by means of a tabulation projected on the screen, where it caught the attention of every conferee. Sir Otto regards the most fundamental feature in safeguarding species as being the difference in objective recognized by political leaders and by conservationists. The difference in the time scale clearly increases with scientific progress, which is another way to notice that the clock moves only onward.

7

The Unwanted

WHETHER OR NOT New Englanders appreciated it, 1973 was a particularly good year for mosquitoes. An abundance of rain left puddles everywhere, just as the wet weather had done in 1972. The intervening winter was so mild that the mosquitoes that multiplied in 1972 had plenty of survivors to mate and lay eggs in 1973 wherever they found small pools. Seldom had there been such an abundance of mosquito wrigglers to nourish small fishes, or of adult mosquitoes to be caught by dragonflies, bats, night-hawks, and swallows. No swallow had difficulty capturing two thousand mosquitoes daily for itself and, by regurgitation, for young swallows in the nest.

Male mosquitoes delicately sustained themselves by sipping plant juices. The females buzzed about in search of red blood as though aware that without this dietary supplement they could lay only five or ten eggs, compared to two hundred after a blood meal. The females settled on frogs; turtles; snakes; birds; and horses, people, and other mammals. People swatted at the fragile insects and assembled in irate groups to argue the pros and cons of mosquito-control programs. Should new drainage ditches be cut into every wetland? What biodegradable pesticide could be

spread by airplane to rid the suburban areas of biting mosquitoes? What good was a mosquito anyway?

We tend to admire those few parts of the world that have
no mosquitoes, not to regard the living communities there
as underprivileged and deprived. A mosquito may be the
most important living thing in the world for another mosquito of the same kind and opposite sex. It may be essential
for the survival of the infective agents of malaria, yellow
fever, encephalitis, and other diseases. But all of these qualify for membership in any list of unwanted organisms—
ways of life we could dispense with and never miss.

A mosquito bite in New England is more an insult than
an ordeal, an irritation rather than a hazard. The whining
sound of the insect's wings during her approach seems a
threat. It ceases as the insect settles, lighter than any feather,
on her victim. For a minute or more, she may make no further move. Then she slides six lancets, finer than hairs and
lubricated by saliva, imperceptibly into the flesh she stands
on. She finds a blood vessel. Her lancets form a tube
through which she sucks liquid nourishment into her swell-
ing, reddening abdomen. All the while, an anesthetic in her
saliva soothes nerve endings near the minute wound she
makes. Out slide her lancets. Heavily the mosquito gets airborne again. She drones away to digest her blood meal and
to transfer nourishment from it to the eggs developing in
her ovary. But no longer may she leave her victim infected
with malaria or yellow fever. In New England she has nowhere to acquire these perils of earlier centuries. The worst
she could do would be to transfer the virus of eastern
equine encephalitis. The chance that she will do so is much
less than one in a billion.

Eastern equine encephalitis is native to the living community east of the Appalachians, all the way from southern
New Hampshire to Florida, and westward around the Gulf
of Mexico to southernmost Texas. But it is a disease of
small perching birds, such as sparrows, and of reptiles and
amphibians. It may well spend the cold months in the

blood of cold-blooded animals. One particular kind of mosquito, the common *Culiceta melanura,* bites all of these animals but shows virtually no interest in the blood of domestic mammals or people. The freakish conditions under which *C. melanura* will bite a horse or a person seem to arise in New England rather than elsewhere in the nation and in Massachusetts least infrequently. Encephalitis was first discovered in a human victim in 1938. It remains rare in people and has not caused a single confirmed fatality since 1956. During recent decades, one person definitely has had the disease but survived it with some impairment, as 40 percent ordinarily do.

The severity of the encephalitis when it does strike people justifies the operation of the Encephalitis Field Station in Lakeville, Massachusetts, under the state's Department of Public Health. Technicians there monitor the summer abundance of *C. melanura* mosquitoes and test blood samples from any person, horse, or bird suspected of bearing the infection. Fortunately, horses can be immunized inexpensively and reliably every year before the mosquitoes start flying. Yet many horse-owners decline to bother, even in years when they are warned that the hazard of encephalitis will be at a peak. In 1973, twenty-one horses that had received no shots were found to harbor the active virus in Massachusetts. Another sixteen horses with no record of shots had antibodies against the disease in their blood, showing that they had recovered from an attack at some earlier date without the illness being recognized.

The 1973 threat to horses and people caused greater apprehension in New Hampshire. When a few ring-necked pheasant at the State Game Farm in Brentwood died and encephalitis was suggested as one possible cause, the governor did not wait for confirmation based on tests. He ordered the extermination of the entire flock. Some twelve thousand birds were promptly gassed to death, as were the golden pheasant, ducks, and geese kept on display at the farm. The governor's drastic decision may have been influenced by a newspaper scare story: a sailor in Portsmouth, New Hampshire, was being tested for possible infection with the dread disease. The extermination program was completed before the sailor's illness proved to be something else and before the laboratory failed to confirm encephalitis in the original pheasant that died. The governor could not get rid of *C. melanura* so suddenly, but he could, and did, cancel all plans to raise more pheasant for release in New Hampshire where they might possibly contract the disease.

Biting mosquitoes and the infection they transmit to small birds and pheasant appear to follow a cycle in abundance, cresting every six or seven years. But every summer, several people telephone us to ask whether it would be wise to establish a mosquito-control program. So far, we have feared the palliative measures (which could never end) more than the mosquito bites that are part of summer in New England. No one wants mosquitoes. Yet few are willing to police the area to rid it of every breeding place that can be found. (Cans, bottles, discarded automobile tires, even gutters along eaves may hold water long enough for a mosquito or two to raise a big family in them.) Still fewer are willing to prohibit house construction on or near swampy land.

More often we fear the arrival of a breeding population of some foreign pest. One that may someday affect a major part of the United States, perhaps the whole area south of a line from New Jersey to Puget Sound, was noticed first in 1918 at the port city of Mobile, Alabama. A small, dark reddish ant from South America, it is called the "fire ant" because of the intense burning sensation it produces on human skin. The quarter-inch insect clings aggressively, pinches up a bit of skin in its jaws, and jabs the taut surface with a particularly venomous stinger. Farmers complain that fire ants build cement-hard mounds as much as three feet in diameter and equally high. Bulldozing them is expensive. The mounds bend or break ordinary agricultural machinery. Any attack by hand seems out of the question because the ants pour out of any damaged nest and attack vigorously. Yet so few Alabamians met this introduced insect that for years little attempt was made to control it. Owners of land with heavy infestations found it easier to abandon the land than to get laborers to work on it. When the area grew up into forest, as it would in a few years, the ants disappeared.

During the 1930s, something happened to the fire ants in Alabama. Either the dark reddish ants mutated into a pale race or, through a most improbable coincidence, a second invasion of fire ants from South America reached the identical U.S. port. The dark fire ants are native to southern Uruguay and northern and central Argentina; those that arrived in Mobile probably boarded a ship at Montevideo or Buenos Aires. The light-colored race is more widespread in South America—in Uruguay, Paraguay, Argentina, Bolivia, the Guianas, and far up the Amazon and elsewhere in Brazil. This is the race that the famous explorer Henry Walter Bates mentioned in his book *The Naturalist on the Amazons* (1863). He told of whole villages from which the human inhabitants fled as soon as fire ants moved in.

The pale-colored fire ants could rout field laborers. These insects would attack newborn pigs, calves, and lambs, sometimes with fatal consequences. They would swarm over helpless young birds in the nest, even entering eggs as soon as the hatching chick had pipped a hole through the shell. Occasionally the ants forsook their normal diet of seeds and caterpillars to eat corn seedlings and those of other important crops or to peel off the bark from young trees, girdling them. Only Alabamians, however, developed enough concern over fire ants by 1937 to begin a program aimed at controlling the pest. Local applications with insecticides had moderate effect. By 1950, entomologists had still found no way to destroy more than about 90 percent of the nests they treated. Yet the problem did not receive a high priority. Fire ants stung fewer people than did familiar bees and wasps. And the new pale race built smaller mounds than the dark reddish ants, which the pale ones soon displaced.

The pale ants spread rapidly. By 1940 they reached Florida to the east and Mississippi to the west. In 1953, entomologists found them in Louisiana. By 1957 they were in Texas, Arkansas, and Georgia. Oklahoma and both Carolinas reported fire ants in 1958, and Virginia in 1959. At first, people hoped that frost would kill the ants each winter. But they burrowed more deeply and stayed alive in unfrozen soil.

The explosive spread of the pale fire ants brought these insects into painful contact with far more people. Even if the insects were primarily a nuisance, demands reached the U.S. Department of Agriculture for emergency measures. Congress allocated $7.2 million for a three-year eradication program extending from Florida to Texas. Both infested and susceptible land totaling between thirty-one thousand and forty-seven thousand square miles were to be sprayed from the air with highly poisonous insecticides (the chlorinated hydrocarbons dieldrin and heptachlor) at the rate of two pounds to the acre. By 1970—the third consecutive year of treatment—fire ants would be no more.

Immediate protest came from the U.S. Fish and Wildlife Service and from sports fishermen. Tests could be cited showing that a single pound of dieldrin to the acre of marshland would destroy the entire fish population of more than thirty species. A pound to the acre killed virtually all larger aquatic animals except a few snails and other shellfish. Everyone agreed that whenever poisons are broadcast from airplanes, some areas are missed, leaving islands from which a pest quickly reinfests adjacent land. Other areas accidentally receive double doses, although the first treatment provides the highest concentration (to get the quickest kill) consistent with safety. No one yet knew the long-term effects of these chemicals on man and domestic animals, let alone on wildlife.

Conservation departments in Alabama, Georgia, Louisiana, and Texas had new information within weeks after the first aerial application of insecticide granules. Fire ants were still visible and active, but equally obvious were "alarming numbers of dead insects, many of them important in insect control or as bird food." Songbirds and game birds lay dead in fields, streets, woods, and marshes. Ornithologists could count only between 3 and 8 percent of the number of kinds of birds seen previously. Squirrels, nutrias, cotton rats, white-footed mice, raccoon, opossums, and armadillos died in large numbers. Rabbits were hard hit or exterminated, and fox cubs were found dead in their dens. The two insecticides were found in the flesh of these animals. Brood sows lost their litters. More than a hundred head of cattle near Climax, Georgia, died suddenly after being sprayed with dieldrin from airplanes.

Water animals proved more susceptible. Fishes, crustaceans, even snails were devastated within a few days after the aerial poisoning of marshes, ponds, and streams. Survivors were still dying seven weeks after the spraying ended. But fire ants remained.

For 1959 the federal program to exterminate fire ants offered a new recommendation—not 2 pounds of dieldrin to the acre, but 1.25 pounds—with no explanation. In Alabama the state legislators held hearings and decided that the fire ant posed too little threat to agriculture or livestock production to warrant further participation. It was better to live with unwanted fire ants than to keep the ecology out of joint by spraying poison over so much of the state. Louisiana declined to vote the matching funds for a continuation of fire-ant control, because insect damage to sugarcane had increased so much wherever the insecticide had been broadcast. Georgians, too, admitted that fire ants were primarily a nuisance. Moreover, the stinging insects kept reappearing in areas that had been declared free of them.

Early in 1960, the Food and Drug Administration of the Department of Health, Education and Welfare discovered that weathering caused heptachlor residues to transform into a persistent and extremely toxic derivative. Traces of it could be detected in milk and meat from animals that fed on pasture or forage crops treated with this particular poison. Further use of heptachlor was ruled out.

The manufacturers and distributors of insecticides did not give up. They experimented for eight years with a suggestion made prior to 1960 by the U.S. Department of Agriculture that the ants be given a poisoned bait to carry home and eat. Limited tests showed a 95 percent kill. This was enough to convince the department to authorize for 1970 the dispersal from aircraft of bait pellets over more than seventeen thousand square miles of the southern states. "Mirex" suddenly became a familiar name for these pellets made from corncob grits and soybean oil, impregnated with chlorinated hydrocarbon insecticide. According to the proponents, 1.25 pounds of pellets to the acre would perform the hoped-for miracle, although only a tenth of an ounce of insecticide would be included. Accumulation would be negligible, they claimed, even with three applications annually. The program would be well worth the $8 million annual cost, since it would be so much more effective than the previous attempts to control fire ants. Few mentioned that the expense to date had already exceeded $100 million out of tax income, with negative rather than positive results.

Less than a year after the initial wide-scale application, the Environmental Defense Fund had enough hard evidence for court action to stop aerial applications of Mirex. Pellets had gone into wetlands and forested watersheds, contaminating runoff. Juvenile shrimp and crabs were dying of the specific poison. Residues of it could be found in the fatty tissue of people who lived in the vicinity. Soon the National Cancer Institute reported tests showing that Mirex could cause cancerous tumors in mice and rats. The Environmental Protection Agency reacted by limiting to one per year the number of permissible aerial applications and prohibiting any use close to waterways, watersheds, or

wetlands. Admittedly this rendered the whole program in-
effective, for the protected lands would surely harbor fire
ants. Each nest can produce hundreds of winged queens
annually. Every queen has the capability to fly several miles
and set up a new colony. The new restrictions led the secretary of agriculture to cancel all efforts aimed at fire ants,
even the less expensive ground-level operations that might
help slow the spread of the infestation. Edward O. Wilson,
the distinguished expert on social insects at Harvard University, hailed the decision as a wise one, much too slow in
coming. The whole program in the South impressed him as
"the Vietnam of entomology. It's time to let the taxpayers
off," he said. He could not know that within seven months,
the makers of Mirex would have devised a new form of bait
pellet to kill fire ants and would be insisting that it receive
an extensive trial.

Fire ants are merely the first of two waves of venomous
insects known to be spreading northward from South
America. The second, a horde of stinging bees, may be
worse, at least in relatively frostfree parts of Florida and
California. It could reach them within a decade. Already
the National Research Council has dispatched a study team
to seek ways to stop the potential invader, although the stinging bees themselves have not yet reached the Isthmus of
Panama. Their evil reputation has flown before them.

The unwanted insects are honeybees of the extremely in-
dustrious African variety. They forage for nectar earlier in
the morning, later in the evening, and in cooler weather
than the European honeybees. Consequently the African
variety produces 30 to 80 percent more honey. These
fierce bees overpower any of the ordinary domestic honeybees, entering their hives, killing the occupants, and taking
over the store of honey as booty—if not the hive as well.
This competitive advantage explains why apiculturalist
Warwick Kerr of São Paulo, Brazil, did not attempt on African soil to cross the African and European races. Instead,
he brought twenty African queens to his laboratory in

South America and mated them with European drones.

The hybrid bees proved at least as intractable as their African ancestors. Still the beekeepers hoped for an improvement. "We thought that when they got acclimatized, they would become civilized," admitted the Brazilian bee expert Father João Oscar Nedel, S.J. "But the exact opposite has happened." The hybrid bees, moreover, showed a disconcerting readiness to swarm repeatedly. In 1957, less than a year after the African queens were imported, 11 hybrid queens escaped with swarms of workers and established themselves in freedom. Each had the potential to produce 450,000 new bees annually and to continue at this rate for four or five years. Father Nedel reacted quickly. He induced the Brazilian Ministry of Agriculture to order all African bees and hybrids exterminated, wherever found. Unfortunately, the ministry did not see fit to follow up this command with any program to indemnify beekeepers for destroying their swarms or to reward them for seeking out escaped colonies. The ruling was never enforced. African bees continued to expand their range about two hundred miles each year. They found nesting sites in hollow trees and under low eaves of buildings. They repelled unskilled people who came near and quickly proved ready to make unprovoked attacks on human beings and livestock alike.

By 1965 the unwanted bees had killed four people in Brazil, plus uncounted horses, mules, pigs, dogs, chickens, and wild birds. The official Brazilian tally of human deaths from stings of African honeybees reached ten in 1972. Already the swarms had spread into tropical rain forests of the Amazon basin, into the pampas of Argentina, over an area the size of the United States east of the Rocky Mountains. Throughout this area they replaced the domestic honeybee. Soon the whole of South America would be included in their range. Could a Maginot Line for fierce bees be established in Panama by the time the African hybrids reach Colombia?

Scientists in Central and North America realize how out

of joint the ecology of these more northern lands will be if the fierce bees succeed in spreading there. Rapidly they would substitute themselves for the familiar bees of the gentler race. A new public health hazard would arise, particularly for those people and domestic animals that develop an allergic sensitivity to even a few bee stings. Any significant increase in the frequency of bee attacks would probably lead to insistent demands for eradication of all honeybees near every populated area. No longer would southern culturists be called upon for replacement queen bees to sustain production in northern hives, nor would beekeepers anywhere be able to supply the market with so much honey or beeswax. The practice of transporting honeybees seasonally by truck might have to be abandoned. At present, many farmers depend on these "migrant" bees to pollinate crops valued at $6 billion annually. A serious decline would follow in production of apples and cranberries in the North, cotton and citrus fruits in the South, seed alfalfa in the Middle West, and other crops in California. Wild kinds of vegetation and then the animals that depend on wild fruits and seeds would suffer too from any limitations imposed on the widespread honeybees.

So far, our vaunted technology offers no way to shut out the unwanted bees without destroying the natural world we long to protect. Technology of the 1960s and 1970s allowed "cosmonauts" to circle the earth in outer space and to explore the moon. We can photograph a missile installation or a parked truck from a "spy-in-the-sky" satellite, detect thermal pollution in clear water from a coastal industry, locate diseased areas in forest trees far from any road, or advise an icebound ship in antarctic waters where open channels are. We can monitor the movements of grizzly bear, elephant, or wolf if it has a signal-transmitting collar or lead a whaling ship to every whale in its vicinity. We gained from these two decades new views of minute details of insect and pollen grain, seen through scanning electron microscopes. We brought microcircuitry to such perfection that hand-held computers came into general use. Technology revealed to us the genetic uniformities underlying the coded heritage of molds and mice and men. Yet it offers no harmless cure for stinging bees and ants.

To discover honeybees that are so ready to take the offensive shocks us. Suddenly we realize that the various environments of our "Spaceship Earth" differ distinctively in that some can tailor a species to be aggressive and others allow the same species to become compatible with mankind. Honeybees may well be the second member of the animal kingdom that our ancestors domesticated, preceded only by the dog. The origin of beekeeping is lost in antiquity. Yet, until quite recently, it was a sedentary operation. Beehives were not moved about week after week. They remained in place while worker bees worked themselves to death and were replaced by others, all sterile daughters of the same impregnated queen.

Present practice is to maintain vigor and productivity within the hive by periodically substituting a new mated queen for an aging one. This can be skillfully accomplished with only a few minutes' interruption in the activities of the worker bees. Now all of this scientific exploitation and art in management seems threatened by a few genes, inherited from the African race of the honey-maker. After nearly eighty centuries, the need to tame the insect arises again, now in near desperation.

We read new meaning into the boast of Panamanians that their country is the "Crossroads of the Americas." People there make a living from the floating commerce of the world, as it glides through the man-made waterway of the canal between the Pacific and Atlantic oceans. Geological forces fortuitously raised the isthmus as land to live on, as a barrier that could require a toll to pass. Can this same country now exact a fee for blocking the movement of hybrid bees along its length, across the canal? Red lights for the insects, green lights for ships?

Actually, the Panamanian isthmus presents stop signs for

other potential pests. One of them—the virus of foot-and-mouth disease—is waiting in Colombia, ready to spread through Central America and Mexico. It is still blocked by those roadless jungles of Darien province in Panama. Disease specialists expect the virus to sneak through, despite surveillance, if ever the natural barrier is breached, as it would be if this last 250-mile link in the Pan-American Highway were completed and opened to traffic.

Now we have added reason for apprehension when the news tells of coming progress on this link. No longer need we merely regret that the unconquered territory of the handsome Choco Indians would be invaded by drivers of big machines. Our sentiments stem from a rainy-sunny Sunday spent at the head of navigation on one river in the Darien, while these indigenous people came from far upstream in long dugouts loaded to the gunwales with plantains and bananas. The Chocos, with light skins painted with complex black designs, maneuvered their craft skillfully to stop beside shabby boats of traders from the outside world, who bartered for the fruit with cloth, fishhooks, beads, and even ornamental shells. Their business finished, the Indians paddled their dugouts upriver with undiminished dignity, each canoe now riding high and showing its decorative markings along the sides. Off they went, into the mysterious upper reaches that were off limits to outsiders, perhaps (so the rumors held) to villages where cannibalism was still practiced upon occasion.

How could we not applaud a court action in 1975, halting all work on the highway into the Darien until the Panamanian Department of Transportation could complete an environmental impact statement on the controversial, expensive project? Fear of a virus accomplished what no consideration for the Choco Indians had been able to do. Foot-and-mouth disease is a debilitating infection of the Old World. Only since 1950 has it become widespread in South America. Its cause is one of the most minute virus agents known. It attacks deer, cattle, sheep, goats, swine, and other animals with cloven hoofs. It can be transmitted by imported livestock, infected materials such as hay or garbage, imported smallpox vaccine, or a wild deer running along a new highway, dodging the customs inspectors and heeding no stop signs. The consequence could be a crushing blow to the large, intermingled populations of deer and livestock in the United States.

Panama stops the spread of another unwanted animal, a poisonous sea snake, just as it has for at least a million years. Few people thought about this fortuitous situation until after 1964, when exploration began in earnest to find an alternate route for a transisthmian canal, one that would be entirely at sea level. Then biologists pointed out that the venomous sea snakes abounded in coastal waters along the entire Pacific side of the isthmus. A sea-level canal might allow the snakes to reach the Caribbean and Atlantic, where no sea snakes of any kind occur. Progressively these reptiles, which possess the same type of poison as cobras, could spread to the West Indies, the whole rim of the Gulf of Mexico, and much of the Florida coast. Although sea snakes are fish-eaters, they sometimes bite people. Scuba divers who are poking about where the snakes seek food are particularly vulnerable. Resort-owners and fishermen see a major loss to their business if the venomous reptiles reach the Atlantic side.

By what magic does the present Panama Canal prevent sea snakes from accompanying ships from the Pacific to the Atlantic Ocean? The canal has been open ever since August 1914. Sea snakes swim occasionally near the locks at the Pacific end, where incoming vessels are raised and outgoing ones lowered. This action is required because the central part of the waterway—Gatun Lake—is maintained artificially at eighty-five feet above sea level.

Sea snakes can tolerate fresh water, but apparently they are too sluggish to swim against the seaward flow of fresh water from the present locks. Routine operation of the canal is based upon this flow. It begins when water from the lake is released to fill the nearest locks, then from lock to lock, and eventually to the oceans. Replenishment is from the Chagras River, which was dammed to produce Gatun Lake. The river, however, flows less vigorously during the dry season. Then the whole watershed that drains into the lake provides barely enough water to operate the present dual set of long, narrow locks. This limitation renders impractical either the enlargement of the present locks to accommodate broader ships or the installation of a third set to cope with increased traffic. The flow that is too much for sea snakes is too little for a more modern waterway.

Engineers see more political obstacles than formidable physical challenges in putting through a sea-level canal elsewhere. It might be built parallel to the present waterway across Panama, or in adjacent Colombia, or nearer North America, perhaps in Nicaragua. The shortest of these various possibilities would be longer than the Panama Canal. Would sea snakes work their way through a new, broader channel? At first, the risk seems great—perhaps too great to take. Surely such a waterway would become salt water from end to end, flowing from the Pacific to the Atlantic side. Do oceanic currents not sustain a difference in sea level of nearly fifteen feet between these two coasts? A wall of water can be visualized, pouring from the Pacific, coursing across the isthmus like a tidal bore in a river. Then an archaeologist pointed out that an aqueduct needs a gradient of fourteen feet to the mile to keep the water flowing—a fact discovered in Roman times. Fifteen feet in fifty miles or more would have no effect, because the channel would exert so much frictional drag on the water.

Scientists from the Battelle Memorial Institute in Columbus, Ohio, investigated the climatic conditions along each proposed route. William E. Martin, who directed the investigations into possible ecological effects, assured us that so much rain falls on these parts of Middle America that any sea-level canal would be kept fresh, outflowing at both ends. Sea snakes would still be repelled, prevented from making an easy transit to the Atlantic side.

The sudden interest in sea snakes, and fears that they

might put out of joint the ecological conditions for mankind along tropical coasts where none now occur, brought
another realization. These venomous reptiles must be relative newcomers along the Pacific side of the Americas. Had
the unwanted swimmers been present a few million years
earlier, they would have encountered no barrier of land.
Until late Pliocene times, the oceans were connected where
Panama now stands. For a while, Central America was an
island bounded on the northwest by the Straits of Tehuantepec and on the southeast by the Straits of Panama. This
explains why the majority of marine creatures on the two
coasts of the present isthmus are so similar. They differ
chiefly in relation to the unlike habitats produced and now
maintained by oceanic currents along these coasts.

Prior to the opening of the Panama Canal, a ship had to
detour far south and risk stormy weather in the Straits of
Magellan, circumnavigating South America to pass between
the Pacific and Atlantic oceans. Given time, sea snakes
might take the same route except for one feature. They
cannot tolerate the cold water in the Humboldt Current
northbound from Antarctica along the west side of South
America. The snakes are also common along the Indian
Ocean coast, but another cold current, the Benguela, prevents the black-and-yellow sea snake from entering the
South Atlantic around the tip of Africa. They range as far
as Cape Agulhas—the southernmost point of the great continent—where the warm water of the eddying Indian
Ocean abruptly meets the cold current.

No one knows how long the unwanted snakes have been
waiting for an opportunity to round the Cape of Good
Hope. Their ancestors inhabited tropical waters around
Malaya and the East Indies, where most of their relatives
live today. Just one species has extended its maritime domain so far and now threatens to go still farther.

For the present, plans for a wider canal across the American barrier between the tropical Pacific and tropical Atlantic await a more urgent need for action. Yet the sea snakes

may soon gain another route to warm coasts where they will be unwelcome. The Egyptian government, which reopened the Suez Canal in 1975, seeks to widen this sea-level waterway through the Red Sea and the desert. Already the canal has admitted fish and crabs from the Indian Ocean to the Mediterranean Sea. Sea snakes might well follow the same route and enter the Atlantic past Gibraltar. Warm currents, such as transported the famous explorer Thor Heyerdahl on his reed raft *Ra* from Morocco to Barbados in 1970, could carry the snakes to American shores in less than a year.

As we think of the world's most unwanted creatures, we realize that another of them, the African blood fluke, still holds a dominant place following a gamble the Egyptians lost in recent years. The Egyptians ignored warnings from parasitologists that any increase in irrigated lands would be self-defeating for them while sanitation is so primitive among peasant farmers. No one could count on discovery of a control measure, let alone cure an infection, for the blood flukes that multiply in hot, wet fields. Infective stages of these parasites pass from human wastes to aquatic snails, and then to people again through the skin of bare feet, legs, hands, arms, and, if the contaminated water is swallowed, the lining of the stomach. Every attempt to break the cycle in tropical wetlands fails, for the flukes themselves evolve immunity to each new treatment devised to restore health in infected people.

The Egyptians insisted on pursuing their impractical dream, on building a high dam at Aswan. When Western financial centers declined to loan millions for the project they lost patience and seized a source of revenue by taking over the Suez Canal in 1956. They collected tolls for eleven years before military actions closed the canal to all traffic. By then Soviet loans of money and advice had the great dam well under way. It was completed in 1971. Within four years Egyptian officials began to accept the fact that all of the predicted gainful consequences had fallen short, while the list of probable losses had been woefully incomplete. Blood flukes, which many Egyptians continue to call "bilharzias," in honor of Theodor Bilharz, who discovered them in their country, had spread through all of the new irrigated area and infected essentially all of the peasant population. No remedy for the disease (properly called "schistosomiasis") has yet been found. The increase in food production yields no surplus for export, to earn foreign exchange. Despite multiple cropping, it suffices merely to maintain at the same low level of nutrition an expanding local population.

Just as predicted, the Nile loses substantial amounts of water by evaporation when it spreads out in Lake Nasser above the dam, under a cloudless sun and dry winds off desert on both sides. Additional water seeps into sandstone and other rocks that have been desiccating for thousands of years. The residual water has a higher concentration of salts after these losses. Still, it must be shared three ways. One fraction goes to irrigate new areas; it suffices for only about half of the land that was supposed to be brought under cultivation. A second fraction goes through the turbines to produce electric power, including some needed for synthesis of nitrogen-containing fertilizer; this too proves disappointing because water can be spared to run only four of the ten turbines that stand ready. The final fraction goes to agricultural areas that have long been Egypt's mainstay. But farmers complain of poisoned crops from salty water. The water quality worsens northward, toward the Mediterranean. Agricultural land in the delta of the Nile is being abandoned because of uncontrollably rising salinity.

As anticipated, the six hundred miles of the Nile below the Aswan Dam no longer show annual flooding. The muddy river has ceased to rise and distribute a fertilizing silt over its gently sloping banks. For the first time, farmers there need to buy and distribute chemical supplements to sustain their crops. They must pay for something the river formerly brought them gratis. Without silt, the river channel erodes. And where Mediterranean sardines used to enter the mouth of the flooding river, supporting a lucrative fishery, the resource is vanishing. It no longer receives a subsidy of dissolved nitrates and phosphates, brought all the way from the lush marshlands of the Sudd.

Egyptian officials now seek Western help to undo what their dam has done. Quietly they admit that the whole Aswan project was a mistake costing $1 billion to $1.5 billion. It was not even worth forcing the relocation of ninety thousand Egyptians and Sudanese in the Nile Valley for four hundred miles upstream from the dam or the great expense of salvaging the immense sculptures at ancient Abu Simbel. The rest of the antiquities simply disappeared from view, along with the vacated villages, the native plants along the river, and any animals that could not find new homes beyond the spreading waters of Lake Nasser. Dambuilders refer to all of these as mere "encumbrances" in the way of progress. Egyptians themselves are now encumbered, in a legal sense, with obligations to the Soviet Union for the cost of the whole enterprise. That debt must be paid, regardless of unwanted blood flukes or how much the high dam put the ecology out of joint.

Personal experience on three trips to Africa (one of them for longer than a dozen weeks) helps us understand the vulnerable situation of people in environments with strong contrasts. We recall our own surprise that the waters are so dangerous on this immense continent where mankind originated. First, we expected to swim in Lake Victoria, almost on the equator, in Uganda. But when we walked across the golf course in front of our hotel at Entebbe, we encountered two obstacles. The first was a column of vicious driver ants, busily hunting for larger animals as prey to overcome with sheer numbers. The second was a sign at the water's edge, neatly lettered in English and Swahili: No Bathing: Bilharzias. We turned back.

Later and elsewhere on the continent, the hazards often doubled. In fresh waters, the message could be Danger:

Bilharzias And Crocodiles. If the water was saline, there might be sharks as well. Since we declined to swim in the Indian Ocean surrounded by a shark net, we also avoided any encounter with sea snakes, for we knew that they were there too among the sharks.

If we try hard, we can think of native animals as exerting some of the selective forces that continue to influence the course of evolution. In Africa long ago these forces contributed to the appearance of primates so versatile in exploiting their environment that they could be considered human. So extraordinary did this one species become that its members grew overconfident. They boasted of being able to conquer nature, to do without any other kind of life that did not contribute in some obvious way to mankind. Who needed useless animals or plants? The world changed at an ever-increasing rate in unwanted ways to the detriment of people as well as nonhuman life. It changed inadvertently with fire ants and recklessly with great dams, because these biased the ecological communities in directions that suited no one.

We have a good chance to notice human errors, because they occur during our lifetimes or recent history. Fewer are the opportunities to notice nature's mistakes and shortcomings. Some of them are preserved in the fossil record. Paleontologists estimate that, out of all the millions of species the earth has known, less than 5 percent—about two million species—survive as contemporaries of mankind. The rest are all extinct. They were ultimate failures, ways of life that did not work indefinitely. No doubt many of them, such as the great dinosaurs, kept the wild community on edge for millennia before they disappeared.

If we focus on the living species, we recognize that they possess a genetic heritage of incredible refinement, adaptive suitability, and ability to change slowly to produce offspring still more fit. How many should be feared as threats to human enterprise? Richard J. Daum, a leading plant-pest control analyst for the U.S. Department of Agriculture, tells us of current work by specialists to assess the situation. Disagreement arises even in picking the ten to one hundred insect pests already inflicting the most damage to crop plants on farms and in forests of just the one country. Each kind may be a prime problem in one area and of minor consequence elsewhere, despite its presence.

A two-part analysis is under way for 185 different foreign weeds, fungal and viral diseases, roundworms, insects, and snails that could cause heavy loss to seventy-one important crops in the states south of Canada. Not one of the 185 is yet established, although it could arrive at any moment. Information and estimates regarding each kind are being kept updated in the memory banks of a computer so that the device can print out a detailed analysis upon command. On the one hand, it combines what is known about the potential pest: its ease of entry; its potential rate of spread; its ability to cause financial damage; its likely impact on the environment, including aesthetic effects. These are assigned a relative ranking on a scale from 1 to 10. On the other hand, the computer takes into account the present area (in millions of acres) devoted to the crop plant, the expectable yield per acre, the market price, and a multiplier to adjust for the number of jobs the crop generates in the economy. As never before, experts seek to combine their scientific knowledge toward a summary of the most unwanted of potential immigrants.

The preparation of an annotated list is certainly feasible. So is keeping it up to date, even if the number of species rises to 250. Yet a major challenge remains: "How can you train the hundreds of inspectors to recognize so many different potential pests?" we ask. "Can any one person learn to identify each particular weed seed, a plant infected with a fungal or viral disease, a specific insect, or a snail egg? How about a squirmy little nematode worm? They all look alike to us! Will the reference guide be a field guide too?"

"No, no!" Daum corrects us. "The inspectors just have to be encouraged to keep alert and to block temporarily the importation of anything suspicious. Experts have to study the samples the inspectors send in and make a quick identification. The annotated lists will help the experts decide what instructions to give concerning the shipment that is held in quarantine."

"Are there enough experts?"

"There are never enough, and that causes delays."

The dictionary reminds us that *quarantine* (from *quaranta*, the Latin word for "forty") literally allots forty days during which the interactions between inspectors and experts can determine what course to take. Almost six weeks during which storage costs accumulate and, for some imports, daily care must be provided. Less than a tenth of a year in which to judge a sample and get the answer back to the inspection station. That decision may affect the welfare of every other living thing within the protected boundary.

A commitment must be made, intelligently or negligently. Often the choice rests upon one person. King Solomon may never have faced so great a responsibility. Charged with safeguarding so much, the solitary individual can feel very much alone. Human dominance of the earth creates these thorny consequences. To keep them from ripping the ecological fabric upon which mankind depends requires an almost perfect union of inherited intelligence and cultural perceptivity.

8

Giving Time

WHAT A RELIEF it was to stand up straight again! For too long we had been hunched over the camera. All that while a gaily fringed sea slug had crept in a spiral, laying her eggs on a smooth rock covered only by the water of a tidal pool. Not until she finished would our photograph show the egg spiral and the animal together.

A voice interrupted us.

"What was so all-fired important in that pool? I watched you two for almost twenty minutes. All that time you hardly moved."

"A sea slug laying her eggs was important to us," we admitted.

Our questioner could have been a character supplied by central casting in Hollywood. Slight. Short gray hair under a long-visored cap. A pea jacket with two brass buttons missing. Faded blue jeans. Rough boots laced with rawhide. Tanned, wrinkled skin, especially on his hands, one of which clutched a homemade walking stick. Since this was the coast of Maine, he probably was the real thing: a retired fisherman or a lobsterman. As it turned out, he was both.

"You must be professors," he remarked, with no rising inflection to make it a question. "Nobody else would know a sea slug if they saw one. Or care a hang."

We grinned and nodded, agreeing with all of these out-spoken thoughts.

"It takes a long time to do things right," one of us admitted, "and there are endless things we'd like to find out. Maybe you could tell us the answer to one question. What kind of a bar gave its name to Bar Harbor?"

"Underneath it's a sand bar. Did you visit it?"

"Not yet. But we plan to."

"Today'd be good. The moon's near full, so the low tide will be lower than usual—a minus tide. The whole bar will be exposed. Used to be an oyster bed, holding down the sand. No oysters now, though. Only dead shells. Blue mussels smothered the oysters a long time back. Then barnacles covered the mussels. Today it's all patches. When barnacles smother mussels, the mussels die and the threads that hold them to the oyster shell let go. Then very small mussels settle, covering the exposed oyster shell. You'll see everything except the tearing loose of the dead mussels. That happens during storms."

Blue mussels form a dense scalp over the dead shells of oysters on the sand bar for which Bar Harbor, Maine, is named. Periodically, acorn barnacles smother and kill the mussels, which then fail to maintain their anchoring strands and are torn away by storms, dooming the barnacles. A fresh colony of blue mussels then builds on the oyster shells, starting the ecological succession afresh. (Photo by Lorus J. and Margery Milne)

"Does anyone harvest the mussels?"

"Just a few. Folks don't like the dark meat, but it makes good chowder."

"Soft-shell clams are better?"

"They sell better," he responded. His face darkened, and he added, "But clams need time to grow. We'll be out of the clam business in a year if we let those people bring their big dredge here and work over our flats."

"Is this something new?"

"It's a big scow, called a Maryland Clam Dredge. It's fitted up to work at high tide over a clam flat. It has water jets to loosen the bottom and cut a trench about eighteen inches deep and nearly three feet wide. Everything from the trench, including the clams, goes up a chute onto a conveyer belt in the scow. Clams of legal size can be picked out. Everything else, including small clams, stays on the belt to the other end of the scow. They're dumped overboard. Loose like that, the little clams are supposed to dig in and grow. But the dumped material is too soft. The tide keeps exposing them and eroding the bottom. After a dredge has been working for a week, it's caught all the clams there'll ever be from that flat."

"Has it been tried in Maine?"

"Not yet it hasn't. But they know we've got 80 percent of the soft-shell clams in the country. They apply every year for permits. Some year they figure to get permits and clean us out. All it needs is a majority in the legislature. Already a majority are inlanders."

"Are inlanders aware that it takes years of pounding waves to make a good clam flat? Or that clams need years in a clam flat to grow to legal size? Or that little clams come only from big clams that nobody digs?"

"That's it exactly!" he agreed. "You can't run a clam flat like a corn field."

We often recall that morsel of wisdom in other connections around the world. Not that we have anything against corn fields, for we would favor a monument to honor the unknown American Indian who first recognized the value in cultivating corn. That person did more to advance *Homo sapiens* than most people for whom monuments are erected. But corn, like wheat, rice, and the potato, is no wild plant. It is a carefully bred one, the seeds of which some-one tucks into prepared soil and then tends until it has kernels of its own at a preferred stage of ripeness. A good crop not only incorporates sunshine and rain, which are free, but invested energy, which people must get and expend. All too frequently today, the amount of food energy in the crop (captured from sunlight) is less than the energy from fossil fuels that must be bought and paid for to produce that crop: to get the hybrid seed and the suitable fertilizer; to plant the seed; to buy and apply chemicals that will "discourage" weeds, insects, and fungus pests; to harvest the crop; to prepare and ship the products to the people or domestic animals that will eat them.

Corn is a special annual plant, inheriting a style of life that ends after a single growing season. By contrast, many of the wild plants and animals into whose lives mankind breaks to gain quick benefit can live many years. Unlike corn, which is an introduced plant with no natural ecological links to any neighbors except mankind, the wild plants and animals interact as a community, sharing benefits in a marvelous meshwork of relationships. Domesticated corn is like civilized mankind in having lost an ability we recognize in every kind of wildlife. It cannot perpetuate itself under the natural conditions of the habitat without a subsidy of materials and energy from other species. It cannot retain its place by adjusting its activities to the local climate, as a cactus plant or a kangaroo rat can in the Arizona desert. It may grow in an Alaskan garden but not compete with tundra lichens or arctic animals, from prolific lemming to polar bear. Not one wild species seeks (or manages) to subsidize a misfit way of life by borrowing energy with which to reorganize the landscape or lay claim to waters from elsewhere. Only mankind behaves in this manner, with no expectation of ever paying back the loan.

Each solitary plant or animal in the wild lives on a strict budget, following inherited guidelines while acquiring and expending energy. Recent research reveals how completely this heritage directs a bumblebee as it buzzes from flower to flower. Activity must be efficient if enough energy is to be obtained from nectar to keep flying and to accumulate a small surplus; the surplus goes as an investment in producing another generation with the same endowment. The life of a plant is equally circumscribed. A whole tree will perish if its inheritance leads it to invest too much energy in producing leaves on low branches after these are in the shade. They become a liability as soon as they fail to gain from the sun more energy than they cost in growth. Nor can the tree declare bankruptcy and start afresh. Its site in the forest goes to a different tree, one with a more appropriate heritage. This is the basis for ecological succession.

These changes take time. They go unnoticed by anyone who is in a hurry. By contrast, the situations that develop as a consequence of human actions may be evident within a year. We have just seen this in our little town, where the state fish-and-game commission decided to install a fish ladder. Salmon, smelt, and other fishes from the sea might use the ladder and then the millpond. To favor these few species, all natural occupants of the pond were sacrificed. Out through the opened gate in the dam, into the salty estuary, went the pond water, the fishes, turtles, and other creatures that made the pond their home. The muddy bottom dried in the summer sun. Freshwater mussels died, their shells gaping in amazing numbers.

We have no idea where the muskrats went when the pond drained. Or the otters that formerly shared the freshwater mussels when not catching fish. The beaver family tried twice to dam the stream and save at least part of the millpond. But dry weather had shrunk the runoff so severely that their efforts proved futile. They took an alternative move, upstream where the river was narrow. There they built two sturdy dams where none had been before in recent history. Their new neighbors in the human community did not appreciate having trees cut. Rather than take the trouble to put wire guards around cherished trunks, they appealed to the fish-and-game commission to trap the interloping beavers—to get rid of them by any means.

Meanwhile, on the bared bottom of the millpond, shorebirds feasted on stranded insects and crustaceans among drying vegetation—plants that had been buoyed up, although submerged. Soon water lilies, pickerel weed, and many cattails died of desiccation. Not until this devastation seemed complete were the gates closed. By then, most human neighbors realized that the pond would never again offer the amenities to life it held a few centuries previously. That was when forests clothed the watershed, the river ran strongly all year, and the soil below the trees had not yet been washed away to become mud in the pond, and the tidal dam was new across the site of the former natural falls.

A few rains helped the stream refill the pond. The current all next year brought thousands of dead water-lily stems to the calm waters. Some of the stems floated, their leaf scars awash, reminding us of watchful crocodiles or alligators much nearer the equator. But the millpond that was to have welcomed fish from the sea offered little welcome to any creature. It needed time, years perhaps, to recover from a summer without water. What power the democratic process awards the fish-and-game commissioners that they can order such a tradeoff of well-established life for the potential good of something new!

For three years no salmon, smelt, or alewife ascended the new fish ladder, which cost so many lives by putting the whole ecology of the millpond out of joint. Until these seagoing fishes do utilize the freshwater reaches above the dam and improve the sport of fishermen, people will be

Resident muskrats and beavers must seek elsewhere for suitable habitat if their pond is drained. Vulnerable on land, the adult of both kinds go up- and downstream searching for a place to start a new life, just as they had to as juveniles, driven by their parents from the site of their birth. (Photo of muskrat by V. B. Scheffer, U.S. Fish and Wildlife Service; of beavers by Lorus J. and Margery Milne)

critical that money collected as fishing-license fees was spent in this particular way. State agents for fish and game, like their federal counterparts, must always face this test of their decisions. What they patently overlook in their programs is the number of citizens with no interest in harvesting either fish or game, who recognize greater values in unmolested wildlife and in the habitats these creatures need. Many species, whose wild world is constricting as human populations swell, suffer from the biasing of the environment to favor a minority—the purchasers of fishing and hunting licenses.

The unstoppable ticking-away of time and the still-unstopped rise in human numbers tend to sharpen the contrasts between these conflicting interests and to thwart any choice of a unifying policy against which to measure each new decision. We expect efficient management, although often this depends upon ruthless simplification, large-scale operations, and increasing use of energy. The rising cost of energy, like the steady depletion of living resources, may yet turn sentiments toward the alternative option: lower efficiency, greater diversity, and small-scale programs to fit each local scene. We can envy the low-energy requirements of wild plants and animals, although we hesitate to emulate them. Yet the nearer mankind can tailor each culture to match these wild ways, the longer our species can lay claim to a future that is acceptable.

Fostering diversity at the expense of efficiency could offer a lasting solution to many a modern conflict. We recognize this along the California coast, where citizens align themselves into special-interest groups regarding plants and animals just offshore. Several property-owners near Monterey expressed their anxiety over the future of the coarse seaweeds that spread out close to the surface in water as much as sixty feet deep.

"Those kelps are the best breakwater we have," said one. "They rise and fall with the tide and damp the force of waves. This calms the water before it can strike the land.

The healthier the kelp forest is, the less erosion we have."

"Do the kelp harvesters take too much?" we asked.

"Not really. They deliberately leave plenty to grow again, and the kelps grow at a prodigious rate. The kelpmen are very careful not to harm the resource because they count on it for crop after crop, to keep their factories supplied. That's the only way they will be able to continue producing alginates for the market. Alginates from the algae, you know, go into cosmetics, ice cream, and many other mixtures to provide smoothness and body to the mix."

"Then what endangers the kelp?"

"Sea urchins, directly. And abalone fishermen, indirectly."

"How's that?"

"We have just a few families in this vicinity who make a living by diving for abalones. Oh, they buy licenses from the fish-and-game commission. Essentially that's the trouble. The fish-and-game officers know where their salaries come from—from licenses. So they tend to forget that sea otters are an endangered species, totally protected. The officer can fail to apprehend or prosecute any of the abalone people who knife or spear a sea otter or who use a rifle with a telescopic sight to shoot a few from some cliff overlooking the kelp bed."

Suddenly we saw the connection. Sea otters eat an occasional abalone, thereby conflicting with abalone hunters. Yet sea otters are needed to eat sea urchins, which feed on kelps and can destroy whole underwater forests. In places where sea otters are numerous, they prevent sea urchins from surviving long enough to grow shells as much as an inch in diameter. Small sea urchins eat only small holes in kelp fronds. Kelps then thrive, nourishing many kinds of small animals that sea otters and other creatures eat, all without disrupting the coastal marine community. Sediments accumulate on the bottom in the quiet shade of kelp fronds, affording a home for still more kinds of life and nourishing other predators. Scuba divers see rock greenlings darting after smaller fishes, all lurking below the kelp fronds. More conspicuous are harbor seals and bald eagles, fish-eaters that haunt Pacific coasts wherever kelp forests are well protected.

"Since 1970," our informant continued, "the number of sea otters has been downgraded about a third. And some of the sea urchins now attain a shell diameter of three inches or more. Their appetites grow in proportion. Just one big urchin can so damage the rootlike holdfast where the kelp is anchored to the bottom that the whole plant tears loose from the sea floor. Waves toss the kelp ashore, where it dies and benefits nothing more than a few flies. The kelp forests thin out. No longer do they shelter the multitude of microscopic plants and small animals that lived there previously. Even the abalones get less to eat. They don't attack the kelp itself but depend upon members of the kelp community for food."

Who would have believed that the presence or absence of the entertaining sea otters could make such a difference? We look for these animals at every opportunity, hoping to see one floating on its back, one leg clinging to a kelp frond as though it were a mooring line, while balancing a flat stone on its chest and smashing a sea urchin on the stone. Yet the sea otter plays an important role by preventing the sea urchins from wrenching the whole ecology out of joint. Recovery afterward is far harder to arrange—as difficult as restoring a shoreline after storm waves have eroded it away. Recovery takes more time, whereas worsening situations tend to appear quickly, sometimes fast enough to reveal what is going wrong.

A good many people along the California coast would like to see the harvesting of abalones prohibited altogether. This view arises from the fact that harvesters of these large and delectable shellfish distract attention from conserving other, more valuable resources. Despite the high price of abalone steak, the total income from it supports few fishermen and is inconsequential by comparison with the value of

The sea otter, whose lustrous fur coat and liking for abalones make it a target animal despite efforts to protect it, serves as a "keystone species" in the ecological community of Pacific and Arctic Ocean kelp forests. No other predator keeps sea urchins from destroying the kelp, which shelters and nourishes directly or indirectly a vast assortment of coastal life. (Photo by Karl W. Kenyon, U.S. Fish and Wildlife Service)

perpetuating the kelp forests, the diverse life they shelter and support, and the coastal properties behind the natural breakwater that they form. Sea otters serve the marine community as a keystone species, holding everything else together. Their value alive exceeds the price that might be gained from their fine pelts or the quick gains that may be taken by controlling sea-otter populations.

If the sea otter is a keystone, certain other living things in the coastal community must be likened to time bombs. Their numbers explode, causing devastation to almost everything close by. Abruptly they cancel out all the delights that everyone enjoys along a seashore. No one and few nonhuman animals can ignore the change. The beach is deserted. The "red tide" has struck.

The color of the inshore waters for which this event is named comes from the presence of as many as 60 million microscopic cells of a single kind swimming in each quart of liquid. Their whirling motion and the whiplashes that produce it have earned the name *dinoflagellates*. Along the

Pacific coast from California to Peru, the cells that cause red tide are chiefly those called *Gonyaulax polyhedra,* whereas those of Atlantic outbreaks are *G. tamarensis* in cool waters and *Gymnodinium breve* farther south. Most Floridians hear of "Jim Breve," because red tides are far commoner than they used to be.

Ordinarily, any quart of surface water from the appropriate ocean will contain a few of these whirling cells. Each is no more than one twenty-five-thousandth of an inch in diameter. Normally they reproduce at a modest rate, using the sun's energy each day for photosynthesis. They provide some nourishment for small drifting animals of many kinds.

These whirling cells underlie the familiar rule that oysters are safe to eat only during months spelled with an *r.* The minute cells, and others that are like them in containing toxic materials, increase in numbers during the *r*-less months from May through August along all coasts of the North Temperate Zone. Oysters gather many of these poisonous motes without suffering obvious harm. Yet people who eat the oysters—particularly raw oysters—get sick and may die. The same oysters, left unmolested, purge themselves in time by eating harmless foods. So why not wait until the oysters are safe to eat?

When the poisonous dinoflagellates "bloom," their populations number hundreds of thousands to the quart. When they number one hundred thousand to the quart, clams and other shellfish accumulate enough of them to become poisonous. At two hundred thousand, the water turns brownish red and finny fishes begin to die. Some are affected by the poison itself. It dissolves in water and acts like strychnine or curare in interfering with normal control of voluntary muscles by nerves. Other fishes simply suffocate at night because in darkness the myriad dinoflagellates use up the dissolved oxygen. By dawn, when the same whirling cells begin photosynthesis and release more oxygen, the fishes are floating belly up, mere flotsam for the waves to cast ashore. Each onshore breeze brings a mist of water droplets from wave action, each droplet loaded with irritating poison. A person chokes for breath at each whiff and soon avoids the shore. So do the seabirds.

The west coast of Florida suffered a catastrophe of this kind during the first four months of 1974. Never before had we seen the waters so littered with dead fishes or such an incredible sampling on the beaches. By choosing our hour for an inspection, we could keep our eyes clear longer and cough less from the toxic mist. The best time was when a breeze blew from land to sea, preferably early in the morning when the stench was minimal. Low tide proved ideal, for then the most recently arrived carcasses are spread out. Yet currents concentrated the floating fishes in many places, depositing them close to the high-tide line.

It was hard to believe that so many stingrays had lived in offshore waters, only to die of the red tide. Littering the beach were massive drums, sheepsheads, several kinds of groupers, thousands of mullet, mangrove snappers, boxfish, puffers, saltwater catfish, slender snake eels, and reef fishes whose identification would challenge an ichthyologist.

At first, sharks fed on the floating fishes. Later, dead sharks were cast ashore: small dogfish and hammerheads to begin with, then larger blacktips, a nurse shark six feet long, and a few gray sharks that, in life, could have menaced people. Each day the assortment changed. Weekdays and weekends, the civic authorities tried to cope with the carcasses. Men in boats collected more than twelve hundred tons of floating fishes from Tampa Bay. Men with bulldozers concealed a greater number of rotting carcasses in beach sand. There the fishes were out of sight but not out of smell!

So much flesh should have been a bonanza for scavengers. But the choking mist was too much even for birds. Even those that probe the wet beach for small prey—the sanderlings, willets, and turnstones—disappeared. Only the bluebottle flies remained. For a week they were few, because ordinarily any fish on shore is disposed of by birds, raccoon, and hermit crabs before blowfly maggots can benefit. Now the maggots feasted, grew, pupated, and emerged as adult flies in less than two weeks. The shoreline buzzed with the beat of bluebottle wings. More maggots. The fish carcasses shrank to stark white bones. The stench dissipated. The red tide had nothing more to kill. Bluebottles disappeared for lack of food for maggots. The inshore waters and the beach regained their normal state. Gulls, wading birds, and people all returned.

No one is quite sure whether the seventh chapter of Exodus describes the first red tide in history. The waters of the lower Nile turned to "blood," the river stank, and the "fish that was in the river died." If so, Moses was the only man who knew the day before when a red tide would appear. He struck the water ahead of time with his staff, to impress the Egyptians with the power of Jehovah. Even that did not induce the pharaoh to free the Israelites.

Until the middle 1950s, mystery surrounded the sudden increase of dinoflagellates to such prodigious numbers. Then a particularly destructive outbreak chanced to occur along the coast of Peru just as an oceanographic expedition from Yale University arrived with sampling equipment of many kinds. All the reference books had led the oceanographers to expect dry, cloudless days and nights. Instead, they encountered rain, which had been falling on the coastal slopes for several months. The harbor of Callao, on the coast eight miles west of Lima, was full of idle fishing boats. Everything with white paint, in fact, was blackening because of hydrogen sulfide from the stinking sea. Seamen called this unusual phenomenon the "Callao Painter." Fishermen muttered about "El Niño," the warm current along the coast that brought this disaster at wide intervals. The last time previously had been in 1925 and, before that, in 1891.

The oceanographers made their measurements and compared this new information from 1953 with records from earlier expeditions. A logical pattern emerged. All would be well along the Peruvian coast as long as the great eddying currents of the South Pacific drifted northward past Callao. The famous Humboldt Current normally brought cold water from Antarctica all the way to the equator, with a wealth of oxygen dissolved and available to sea animals. At the same time, the current forced up along the Peruvian coast equally cold water from the sea floor rich in phosphates and nitrates. The dissolved nutrients sustain an incredible number of microscopic green plants, especially diatoms, in surface waters. These nourish a dense population of small animals. Little crustaceans become food for some fishes, while others strain out the clustered algae directly. Larger fishes eat the small ones, many of which are anchovettas. Together the fishes support a fishing industry as well as the most prosperous colonies of seabirds in the world. Guanay cormorants, blue-footed boobies, and brown pelicans feast on the fish and then deposit their wastes mostly on the islands where the birds rest and nest. Currently the annual accumulation totals about 350,000 tons on these Peruvian sites. It is harvested as a national resource in seasons when the birds are not nesting and then sold as fertilizer to European farmers. Little is used locally.

About once in seven years, some vagary of the winds alters the flow of the Humboldt Current and suspends the upwelling of dissolved nutrients. The algae reproduce less vigorously. The crustaceans and other small animals become few. Many of the fishes move to cooler depths or farther from the coast. The guanay birds must search ever greater distances for food. Peruvian fishermen hang up their nets, unable to catch enough to justify their time and effort. Even this may be temporary. The men joke about it, saying in Spanish the equivalent of "Six years shalt thou labor, and on the seventh rest."

Sometimes the situation changes. In 1925 and 1953 at least, the Humboldt Current turned westward from the coast near the border between Peru and Chile. A surface flow of warm water from the Gulf of Panama flowed south, unopposed. It began so soon after Christmas that it earned the name *El Niño*—"the Child." Although only about a hundred feet thick, this warm current contributed moisture generously into the onshore winds, as the cold waters of the Humboldt Current never do. Heavy rain began on land. Rivers that had been dry for many years suddenly gushed seaward with great loads of sediment. The silty discharge from the continent changed the chemical nature of off-shore waters at the same time that the salinity decreased. This combination triggered a change: the dinoflagellates bloomed.

Fish died in vast numbers, ruining the enterprise of fishermen. Guanay birds—especially the cormorants, which pursue fishes at the surface—succumbed from poison if they ate the feebly swimming fish and from starvation if they could find no fishes to eat. Confused flocks of the guano-producers dispersed up and down the coast and even inland, where the birds tried to capture insects as substitutes for fish in rain-drenched pastures. Peruvian ornithologists estimated that ten thousand cormorants and an equal number of boobies and pelicans perished, all in a few weeks as a result of the red tide (and El Niño) in 1953.

Now the scientists have a formula that links the appearance of the red tide to specific changes in temperature, in salinity, and in concentration of dissolved nutrients, as these are produced by climatic variations. It fitted the conditions when El Niño returned in 1960 and 1965. Right on schedule in 1972, a new and more severe El Niño followed the familiar pattern. Yet something was different, for the small fishes failed to renew themselves when the red tides faded away. The Peruvian government clamped a limit on fishing, suspecting that too many anchovies and anchovettas had been caught. Where were the one-year-old females, each four inches long and capable of producing ten thousand eggs? Where were the two-year-olds, two inches longer, ready to release a similar number of reproductive units? Neither the major spawning in August and September (the southern winter) or the lesser spawning in January and February (midsummer) provided the expected contributions. With so few of the small fishes in the cold coastal current, the guanay birds went hungry. So did the commercial tuna. So did many of the fishermen, for in 1973 and 1974 they could harvest barely half as much as in 1970 and 1971 for conversion into salable fish meal.

Most of the product from the fish-meal industry, which began in earnest about 1957, goes to European and North American markets. It is used as supplementary protein, added to food for livestock and poultry. These affluent markets adjust to one poor year in seven, when El Niño strikes. But if the supply does not resume promptly, a substitute must be found. Thus, soybeans, chiefly from farms in the United States, took the place of fish meal from Peru in 1973 and 1974. Demand for soybeans skyrocketed. The price quadrupled. Farmers increased soybean acreage 25 percent. Livestock fattened on expensive soybean meal commanded higher prices. When controls were applied to meat at the wholesale level, livestock raisers reduced their production—and simultaneously their need for either soybeans or fish meal. Costs stabilized at new levels.

Now the anchovies and anchovettas have mysteriously reappeared. The guanay birds and tuna are flourishing. But the market for fish meal continues to be a casualty. Almost no one, it seems, wants to rely on unpredictable small fishes. Only the birds, the large fishes, and the guano industry stand to gain.

"What point is there in any big investment to head off red tide?" a Peruvian fisheries officer asked us. His countrymen have considered preventive measures along the coast. Immense reservoirs could be built to hold the runoff from rivers that ordinarily are dry. (Archaeologists maintain that during Inca times or earlier, the same rivers supplied irrigated fields in lowlands that now are too arid for

agriculture and have been since before the Spanish conquistadores arrived.) Yet the investment of so much, if primarily to control nature at the final step after the arrival of
El Niño, might not be sound economics if the disaster
strikes fewer than fifteen times each century. Nor can anyone be sure that a red tide would not recur more frequently
if the soluble materials that flush from the rivers so infrequently were doled out year after year from behind
huge dams. The ecology of offshore waters along the Peruvian coast might then be kept continuously out of joint.
The tuna fishermen and the guano harvesters might be out
of business.

Already people who live and make a living along coasts
of central and southern California, the Gulf of Mexico, and
the Atlantic Ocean as far north as Maine may be experienc-
ing what Peru would if all the dams were built. No longer
can these citizens of the United States shrug off an occasional red tide or count on a wide interval of normal conditions before the dinoflagellates act up again. Disaster can
come every few years, with near disasters threatening in between. The U.S. Public Health Service now monitors
through local laboratories the toxicity level in several kinds
of shellfish. Harvesting is forbidden as soon as more than
eighty micrograms of PSP (for "paralytic shellfish poison")
are detected in each hundred grams of meat. No technique
has been invented to prevent it from changing the coastal
world.

The major rivers of America have dams for flood control
and other purposes. The water from them carries to the
oceans far more dissolved material than it did a decade, a
century, or four centuries ago. Along with chemical compounds from farms, factories, and municipalities goes
waste heat from human enterprises. All of these may pre-
dispose the marine environment toward outbreaks of dinoflagellates. The only extras that may be needed seem to be
heavy rains (which lower the salinity of coastal waters) fol-

lowed by a week or more of calm, bright, sunny days to heat the sea further.

A few people admit to benefits from red tides. In the Gulf of Mexico, at least, the whirling cells that create such disaster for fishes, fish-eaters, and waterfront rental agents may be just the right food for juvenile crustaceans. Edible shrimp travel at maturity beyond the normal reach of red tides, to spawn and die as much as forty miles offshore, where dinoflagellates are relatively few. Gradually the hatchlings work their way back toward coastal bays, growing in proportion to the amount of minute drifting plants they find for food. The young shrimp grow faster when they encounter waters with a sizable number of dinoflagellates so long as the oxygen supply is not depleted every night. After a red tide along the shore has subsided, shrimp appear in estuaries and lagoons. As early as June, July, or August they have developed the robust body size normally seen in October. Fishermen regard these shrimp as perfect for bait—if only there were fishes left for the fishermen to catch. Certainly some fishes survive or move in, find the shrimp, and benefit too. After a while, the fishermen are happy again.

We found one commercial fisherman who could ignore the red tide altogether. He lived a few blocks from the Gulf of Mexico, where neither the toxic mist nor the stench of decaying fish could be detected. His specialty was harvesting blue crabs, and these tended to appear ahead of schedule in any year with an outbreak of the dinoflagellates. His unconcern that the ecology was otherwise out of joint reminded us of the Californians for whom abalones provide an income. Possibly all of us tend to concentrate our interests and to let everything else go out of focus. How often does this lead us to overlook signals of significance and the welfare of our neighbors, of other forms of life, and perhaps of the future too?

Many of our young friends aspire to lives that benefit

from personal commitment, ingenuity, and versatility. They seek a modern place for more generalists and fewer specialists. They vote for diversity and a place in the sun for each minority. This makes them reactionaries in a world that has aimed at efficiency as a special goal. Efficiency means automated production lines rather than cottage industry. It depends upon sacrifices in life-styles, but not in use of energy. The opposite direction seems to offer a longer future for the environment.

Fortunately, specialization in mankind is a cultural attribute. Any excess is subject to quick change and differs thereby from the inherited specialties we see in nonhuman life. We can change—indeed, must change before the year 2000, when the momentum of our past ways will bring the human population to at least 7 billion, all needing food, space, and protection from inevitable pollution. We can take heart from the fact that the anatomical and physiological features of mankind are mostly those of a generalist. *Homo sapiens* is a versatile being, able to progress in any of several directions, chosen according to the values they offer. The future of our own species and every other now seems bound up in the values we choose to recognize.

Today's news is mixed, but almost all of it relates to short-term benefits for the human species and long-term losses for most other kinds of life. Our small town provided us a pastoral sample this past spring. Members of local churches decided to pool resources and provide three dozen small apartments for elder citizens. The site chosen is church property that has been kept as open space, with a fine view of the undeveloped millpond. For years that land has provided a summer bounty of clover for honeybees and cottontails, nest sites for meadowlarks and bobolinks, caterpillars and other insects for goldfinches and indigo buntings, wild strawberries and white daisies for uncounted youngsters. We knew that all of this had to go. We wondered, in fact, where about a thousand voles (meadow mice) from the four-acre field could find new homes when the weedy vegetation that had furnished them with privacy and food was bulldozed out of existence.

In our New England village, where everyone shares gardening experience, Joseph Marelli makes us sensitive to voles. He describes what happens to him: "My neighbor mows the field that bounds my vegetable patch. Promptly his mice move in and begin eating the plants I'm raising. They're even attacking my cucumbers!" Where else, we ask, could the voles get water? Their normal diet has been sundried to cut hay. The voles have had to switch to emergency rations, not to favored foods. From the church property too, the voles might make their new needs known in some unexpected way. Yet a bulldozer would affect them more than any mowing machine. Possibly most of the voles perished as the steel blade shoved the topsoil into huge mounds to save its organic matter. One rarely imagines dead organic matter being made from live mice on so large a scale.

We did seek out the foreman on the project and drew his attention to a venerable pussy willow bush. It grew in a damp spot at the edge of the road. Each spring, townspeople came to that bush to cut withes, to see their catkins open indoors. It reassured them to know that the green world was ready for another summer.

The foreman thanked us for telling him about the pussy willow. He assured us that the bush would not be harmed. "It's on town property, anyway," he said. "We would be liable to a $5,000 fine if we destroyed that shrub. I'll have some bales of straw put in a semicircle upslope, to make it conspicuous. That will also prevent any mud from spring rains from collecting around the pussy willow roots, now that we're baring the earth higher up."

For weeks everyone worked around the bush, respecting it and the bales of straw. No earth was pushed against the shrub. The backhoe operator never broke a single branch as he cut a deep trench close by with his machine. In due course the pussies opened, offering golden pollen to wind and bees, and delighting the passersby. Fresh green leaves expanded, as in previous years. The bush seemed safe. Its future looked good, even though we doubted that the indigo buntings would use it again as a place of concealment for their nest. Perhaps they would return to it another year, when all the commotion of building would have ended.

Then came the day when a new operation began: cutting a crosswise trench for a drain pipe under the road as a culvert. Its high end would be a cement manhole covered with a grating, a few yards away from the pussy willow. Between eight in the morning and five in the afternoon, the job was done. A bulldozer smoothed the earth. But every trace of the bush had vanished, along with those bales of straw. What evidence could anyone offer that the willow had ever been there, stood on town land, or been worth a penalty of $5,000? Oh, well. It was only a pussy willow. There must be another one around somewhere.

This faith in continuity "somewhere" has applied to many species. It may have steadied the aim of the gunner on Barbados who shot the last known Eskimo curlew alive. It may have strengthened the arm of the person who clubbed to death for sport the last dodo on Mauritius in 1693. Somehow this one turkey-sized, flightless pigeon became the symbol for exterminated wildlife.

Now it is clear that the same island in the Indian Ocean lost two other kinds of birds, the broad-billed parrot and the Mauritian rail, in the same century and another, the blue pigeon, in the next. Currently the islanders are aware that two more members of their very special bird list are endangered. One, the Mauritian kestrel, is the rarest bird alive: in 1964, an estimated ten to twenty survived, and in 1975, just two mated pairs and a solitary adult remained free, all of them in the Black River Gorge. The gorge was declared a national park, as though these three little hawks would venture nowhere else in search of food. The whole island has only 174 square miles. One more pair of the kestrels is captive, bringing the total for the species to just seven individuals. In 1974 the captives produced three eggs, closely supervised by the International Council for Bird Preservation, the New York Zoological Society, and the World Wildlife Fund. One egg produced a chick, but it died a few days later when the incubator malfunctioned. The other eggs failed to hatch. Can the world count on better luck or better management? The birds can count only on inherited behavior in one another. This has not given them much margin of safety since mankind arrived on Mauritius in 1598.

We think of the last pair of great auks (garefowl) on earth. They were shot by a fisherman who knew that none of these flightless, goose-sized birds had even been sighted from any European coast for a decade. To get their skins for sale, he climbed bleak Eldey Island, a volcanic upthrust some twenty miles southwest of the Icelandic coast, on June 3, 1844. His effort earned him £9 sterling and ended the possibility that the species could continue. Only one lone great auk was seen later—in 1852, swimming after fish, as millions of its kind had formerly done, over the Grand Banks off Newfoundland. What happened to the millions? The Indians ate some and used the beaks for arrow tips. Europeans killed the rest, some for food and bait but most to get feathers cheaply for manufacturers of feather beds. The last great auks became valuable trophies simply because soon there might be no more to kill.

This attitude continues to beset wild things. Not long ago we pursued a rumor, then neatly lettered small signs beside the road, to witness a phenomenon in Scotland. We signed our names and walked from the parking lot along a covered arborway for about a quarter of a mile. At the end of this camouflaged approach stood an inconspicuous hut. It shielded from any weather several tripods that supported spotting telescopes or high-powered binoculars. All aimed at the same nest, high in a distant tree, where a pair of ospreys had young to rear. These birds were special—the only ones in all of Britain—for native ospreys had been exterminated there years before. Probably the pair had come

Ospreys are again raising families of young and increasing in numbers, now that the concentration of toxic pesticides is diminishing in the environment of the fishes upon which these alert birds feed. (Photo by William C. Krantz, U.S. Fish and Wildlife Service)

from Scandinavia across the North Sea. Scottish ornithologists wanted to be sure they would stay and succeed. The contributions we dropped in the collection box would help to pay for an honor guard, including determined people who sat or slept concealed right at the foot of the nest tree. If anyone tried to interfere with the osprey nest, these guards were prepared to resist. "Are they armed?" we asked. "Not with guns," came the evasive reply. "With brass knuckles, then, or something similar?" we persisted. A confirming grin was the only response.

In North America we see that the osprey population is on the rise all along the Atlantic coast. Improved success in reproduction by these birds implies that they are no longer accumulating so much DDT and other kindred chemicals from the fish they eat. Their eggshells are thicker and sturdier, their young more normal in behavior. We read with equal satisfaction that the concentration of pesticide residues in songbirds is declining too. Yet still more time must be allowed before everyone who might profit by poisoning the environment abandons this practice.

The continued applications of endrin and other chlorinated hydrocarbons in the great Mississippi basin dashed hopes that brown pelicans could again patrol the waters of Louisiana, where the pelican is the state bird. Each year since 1969, state officials have been trying to reintroduce pelicans where pesticides killed them off completely. Pelicans with less contamination were brought from Florida and freed just west of the mouth of the Mississippi River. Enough of them reproduced for their offspring to offset natural deaths. A population of about 450 gave many a person reason to believe the situation was improving. Then in 1975, about 80 percent of the 450 pelicans in Louisiana waters died. The rest appeared doomed. Every individual that was tested contained a lethal dose of endrin in its brain, plus significant poisoning from seven other pesticides.

Apparently many agriculturalists, foresters, and mosquito-control people remain unconvinced by Rachel Carson's message in *Silent Spring* or by the multitude of proven cases extending the evidence in subsequent years. Our hope for an environment free of poisons must await further evolution of immunity in the insects the poisons are designed to kill. Fortunately for wildlife, this evolution in most areas has been quick. Now ornithologists from Cornell University, in cooperation with the U.S. Fish and Wildlife Service and the U.S. Army, are releasing young peregrine falcons at the Aberdeen Proving Ground in Maryland and other sites. The hand-raised birds of prey are intended to reintroduce these swift predators where DDT brought them to extinction. At the same time, the Wildlife Service itself is claiming some success with substituting bald eagle eggs from nests in Minnesota for the contaminated eggs of bald eagles in Maine. The aging eastern eagles have a better chance to raise replacing broods of young imprinted with a Maine address than if their own thin-shelled eggs were left for them to brood. Sometimes the Minnesota

With a wingspan of almost seven feet, a brown pelican glides in for a landing, using the webbing between its toes as an air brake. The bird stands securely on branches of the coastal mangrove trees in which it nests. But all too often in recent years, pesticides in the fishes with which the pelican nourishes itself and its young prevent production of normal eggshells and chicks. (Photo of flying pelican by Jack Dermid, perched bird by A. Wetmore, courtesy of U.S. Fish and Wildlife Service)

eagles respond to being robbed by laying a second set of eggs. At least as often, they lose a year. Only the future will reveal whether the conspicuous program does more than share the country's growing poverty in wildlife, as an exercise in public relations.

A young prairie falcon has become a rare bird, for this species is now endangered over most of its range in the American West and Mexico. Disappearance of suitable nest sites, prevalence of poisoned bait set out for other predators, and a progressive decrease in natural food combine to reduce reproductive success. (Photo by Lorus J. and Margery Milne, in Jackson Hole, Wyoming)

Many a modern move combines a publicity angle and a concealed explanation. Both appear in news that the Wildlife Service has prevailed upon the Air Force to safeguard the popular whooping cranes. No longer will Matagorda Island off the Texas coast be used as a practice bombing range, as it was from 1942 to 1974. Nearby on the mainland is the Aransas National Wildlife Refuge, where all of the wild whooping cranes in the world now spend the winter. "Cranes Will Get Air Force Island," declared the release from Washington. The detailed account that followed admitted that the gulf side of Matagorda, which is five miles long, could still be leased profitably by the state of Texas for recreational uses, while the endangered cranes would be free to visit the bay side. The state land commissioner favored development of all the island beaches, because they are "the prettiest this side of the Riviera." What have whooping cranes ever done for anyone? This disposition of the four thousand acres of federal land on the island could not remain uncontested. The Texas parks and wildlife commissioner applied to have the whole area earmarked as a hunting preserve for any citizen who bought a license. The U.S. Navy proposed instead that Matagorda be sequestered from public use and cranes as well to become a practice landing site for aircraft stationed at Corpus Christi.

Whooping cranes certainly need wilderness both at the

The plight of the last surviving whooping cranes has become a topic of national interest. One migratory flock makes the hazardous round trip each year between nesting areas in Canada near Great Slave Lake and wintering grounds along the Texas coast. A few captives have produced equally few young under protective supervision. Shown here are a parent with eggs on a nest (photo by Charles A. Keefer), a parent with chick among back feathers (photo by W. F. Kubichek), a twenty-eight-day chick hatched experimentally in Maryland (photo by Rex G. Schmidt), and adult birds settling a territorial dispute while young cranes and Canada geese look on (photo by Luther C. Goldman, all U.S. Fish and Wildlife Service).

southern end of their perilous migration route and in their
nesting territories near Great Slave Lake in northern Canada. These majestic birds, like so many other kinds of native life, can attend in a normal way to the survival requirements of their species only if undistracted by any human
presence. Already their plight is marginal, brought about
by exclusion from all but two fringe areas of their formerly
extensive range. Perpetuity for the whoopers depends
upon freedom from further encroachment, whether by
army bombers, navy planes, helicopters, power boats on the
strait between the refuge and Matagorda Island, vehicles of
any kind on land, or even people on foot. Any crane "is
wildness incarnate," wrote Aldo Leopold. "But all conservation of wildness is self-defeating, for to cherish we must see
and fondle, and when enough have seen and fondled,
there is no wilderness left to cherish."

Reluctantly we must admit that the time has passed when
anyone could think of any place on earth as uninfluenced
by mankind. Far more frequent are localities in which the
human species is so separated from wilderness that the
basic importance of wild plants and animals can be overlooked. We may preface a meal with a moment's prayerful
thanks for meat without mentioning the essential interme-
diaries of sun, green plants, and live animal and butcher.
All of them stand between God and us. Rare indeed is a
thought at bedtime in gratitude for the oxygen we breathe.
We have it by courtesy of green plants. A third of the
world's oxygen production comes from minute drifting
plants in bodies of water—places we treat as the ultimate
sink for wastes of every kind. Less than a tenth of the oxygen supply is from crop plants and forest trees that our
culture admits are useful—valuable economically to mankind. The rest of the oxygen is an unthinking contribution
from weeds and "useless" vegetation on areas of land that
have not yet been paved, poisoned, or built upon. If any
change in the environment were to block the process of
photosynthesis by green plants, the nongreen plants and

animals of the world could convert the oxygen of the atmosphere to useless carbon dioxide in less than twenty years.

Ever since the pathways of carbon dioxide into, and out of, living things were first traced, scientists have marveled at the even balance in the natural system. Yet so elusive are instances of anything resembling a steady state that belief in the balance of nature tends to fade. Is it an illusion? Now philosophers are joining ecologists, pooling their talents toward defining the balance as normal. Recently we heard David O. Edge, who directs the science-studies program at the University of Edinburgh, analyze this concept. He called the balance of nature an unstable equilibrium, a system that has the possibility of being ongoing, recycling its nutrients, allowing energy to flow through. Output and input are matched. Above all, it can evolve slowly.

Edge identified as totally unethical any action of mankind that would bias the balance so far that it could not recover. He thanked ecologists, in fact, for providing a clear picture of normality without human interference. This should serve as a yardstick from which ethical judgments can be made. He looked to other people, with a different background in our culture, to present the legal side and to influence value systems in such a way that the normal situation would continue its uneasy balance. He devoutly hoped there would be months and years enough for sapient citizens everywhere to recognize and react to this modern need.

Mankind began biasing the balance of nature for human benefit about ten thousand years ago by beginning the deliberate cultivation of particular plants and the care of domesticated animals. For the first time, hope arose that a part of nature could be controlled. This initiated a new point of view: the human species was separate from nature, superior, and capable of controlling *all* of nature. Areas where nature was controlled (barring catastrophic weather and the like) would henceforth be regarded with a proprietary interest—that is, as property. All other areas were wilderness, waiting to be subdued. In wilderness, a person was outside the territory that had been made safe. It might be better to destroy all wilderness than to have it threaten the dream of ultimate control.

Homo sapiens did not actually move out of the natural sphere until just over three thousand years ago. About the time of King Solomon, men discovered how to smelt iron. Prior to that time, our species fitted fairly well into the energy pattern that had sufficed for nonhuman life for millions of years. (It still suffices for plants and nonhuman animals. Annually the green plants capture from the sun about 840 trillion kilowatts of power. This energy is released and lost to space at approximately the same rate by the green and nongreen plants, the wild and domesticated animals, and primitive people.) Increasingly, people began using the energy stored in wood as fuel for iron smelters. Supplies of wood grew scarce. Then, in the middle of the eighteenth century, Europeans learned how to advance civilization still faster by use of fossil fuels.

Today we recognize the dependence of culture upon the energy being utilized. Although we still rely for oxygen upon undomesticated plants, the food of civilized people comes from modern agriculture—a technique that employs sunlight to change fossil fuels into edible crops. In 1973 mankind used 70 trillion kilowatt hours of power throughout the world, according to the report of the Energy Policy Project, sponsored by the Ford Foundation. Our one species managed at least a twelfth as much energy as all other kinds of life combined; most of this energy came from resources that are not being renewed. They are outside the balance of nature and bias it severely. The ultimate question may be how much civilization will be left if mankind chooses (or is forced by ecology out of joint) to return to dependence upon solar energy. Can we now resume a place—and not a dominant one—in nature? As Ralph Waldo Emerson said, "This time like all times is a very good one if we but know what to do with it."

9

Why Fight Nature?

ONE OF THE FEW REAL ADVANTAGES we can recognize in being human is that unlike most other kinds of
animal life, we do not have to go searching every year for a
suitable place in which to raise a family. Snug in our
homes, we can watch the resident mammals and birds seek
hideaways, the arriving migrants compete for nest territories or acceptable holes. What an enormous effort goes
into digging a burrow, building a nest, or even refurbishing
an old one after every winter!

By March in our part of America, the male redwings arrive as harbingers of spring. Each glossy male displays his
scarlet-and-golden epaulets and emits his distinctive cry.
He identifies an area of marshland and a high perch that
can serve as a command post, despite the fact that the cattails from the previous year may still be crushed under the
remains of winter ice and snow. By bluff and bravado, or
by battle if necessary, he establishes himself to the exclusion
of other males. The vocal commotion informs our ears of
the time of year, generally prior to the official beginning of
the season. Not for a fortnight will the inconspicuous female redwings come north. Many weeks go by before it is
time to construct a nest in which to tend eggs and young.

The females have abundant opportunities to go comparative shopping among the male redwings and their territories, to establish pair bonds, and to ensure cooperative parenthood for the benefit of another family of young redwings.

During those weeks the grackles arrive. Already paired off, they fly arrogantly into the tops of the white spruces we planted years ago. Iridescent and white-eyed in their maturity, these migrants from the Southeast choose the particular tree in which to construct a sturdy incubation platform and rear this year's young. Not before July ends will their parental duties cease.

One more dark member of the same bird family is due prior to mid-April: the brown-headed cowbirds, which to us are the black sheep of the clan. Brown-headed males and mouse-gray females consort in the same creaking flocks. We resent each one of these birds because its mere existence implies one fewer of the various warblers and sparrows whose nests the cowbirds parasitize. We will not willingly exchange many species for just one of any kind. Properly, the cowbirds are bison birds, native members of the former community on the short-grass prairies. They spread into other parts of America as bison were replaced by cattle and pastures appeared in place of tall grasses and unbroken forests.

The cowbirds' range and numbers are still increasing, as are those of another "blackbird," the introduced European starling. Starlings in our area begin house hunting in late winter, as though determined to occupy everything available before the more migratory hole-nesters reach our latitude. Starlings will even oust a native woodpecker from the nest cavity it has just chipped out of a dead tree.

All of the starlings in America descend from birds introduced from England. Eugene Scheifflin, a wealthy drug manufacturer, released forty pairs of them in New York City's Central Park on March 6, 1890. He freed forty more pairs on April 25, 1891. His literary tastes led him to wish to see in the New World all of the birds mentioned in Shakespeare's plays. And in *King Henry IV, Part 1,* Hotspur says, "I'll have a starling shall be taught to speak / Nothing but Mortimer, and give it to him, / To keep his anger still in motion" (act I, scene 3). The European starling has since spread to Kentucky, California, Alaska, and Bermuda. Everywhere they found food in urban garbage dumps and among short-cut grass.

These four blackbirds confront the farmer and the human residents of cities and military posts less during the nesting season than when it ends. Then perhaps 350 million birds of the several kinds begin to move south. They seek locations within a two-hour flight distance where they can find roosting sites above the ground before dark and seeds or fruits on the ground after dawn. Some individuals find this attractive combination in urban and suburban areas where seeds and raisins are offered regularly at bird-feeders.

The greatest confrontation between blackbirds and people occurs annually in the wedge of states south of the province of Ontario and the Great Lakes. For thousands of years, redwings and grackles have been funneling into this area. Probably they began this routine soon after the retreating glaciers exposed the land and the nesting regions farther north. The birds wing their way from a wilderness that extends to the arctic tundra, to wait out the cold months in the area from eastern Iowa to western Pennsylvania, south through Kentucky as far as Tennessee. There the birds now meet agriculturalists who choose to plant corn and grains or to keep feed troughs full of similar nourishment for the benefit of livestock. The farmers may be armed with pitchforks and shotguns, but they cannot be everywhere at once. The birds are far more mobile. And a flock of black fliers swooping down on domestic animals is enough to drive them from the feed trough. The squabbling birds take over from steer, sheep, pig, or chicken. Every year an increased number survives the winter, which otherwise would be a season of near starvation. Consequently, still more return the next autumn with young to intensify the confrontation.

Damage claims exceed $5 million annually in Ohio alone. An equal amount of loss is evident in western Kentucky. Farmers around Lawrenceville, Illinois, had no difficulty in raising a $14,000 "war chest" toward control measures while their applications for a permit to destroy blackbirds awaited federal action.

The courts approved a proposed action first in Kentucky. Officials in Paducah responded promptly. A second attack came a few days afterward in late February 1975, pressed by five-ton Huey helicopters of the 101st Airborne from the U.S. Army base at Fort Campbell, just south of Hopkinsville. The targets were blackbirds—millions of them. Fully a third were redwings, and another third starlings; grackles and cowbirds made up the remainder.

The control measure appealed to many people because of its simplicity. It would be applied in darkness, after the birds had assembled side by side for the night on the branches of roost trees. The date would be chosen when the weather forecasters predicted night temperature chilling below forty-five degrees Fahrenheit. Helicopters and crop-dusting airplanes would rise to douse the birds with a strong, biodegradable detergent called Tergitol S-9. It would wash the oil from the blackbirds' feathers, making them wettable. At the next pass, the airborne equipment would spray plain water, soaking each bird but harming nothing else. The blackbird would lose heat through its wet feathers. By dawn, each wet bird could die of cold.

Paducah claimed three hundred thousand feathered casualties in a single night, as an estimated 20 percent success. The army killed five hundred thousand, but nine times as many survived to continue roosting on the military base. Still needing attention were almost ten million blackbirds at the arsenal in nearby Milan, Tennessee.

Townsfolk and farmers in five states followed this news

with personal interest. These people could suffer lung damage due to bird-borne histoplasmosis, which is an infection from fungal spores in bird droppings. Crops could be ruined or baby pigs could die because of susceptibility to another disease carried by the birds. Conservationists in other states and many countries marveled more remotely, particularly if they had not read the detailed arguments and impact statements, which were prepared at a cost of over $20,000. These satisfied the Council on Environmental Quality, the Humane Society of the United States, a U.S. district court, and a court of appeals. The National Audubon Society agreed that control was necessary and the method a reasonable one.

Through sheer numbers, the blackbirds won the war in 1975. Millions of survivors dispersed in March, as has so long been their inherited custom. The four species sorted themselves out according to their requirements in nest sites. Thereby they ceased merely to be blackbirds. They became prospective parents and a source of future confrontations with mankind. The birds could not change their ways so quickly. Nor were human residents in the wintering areas ready to yield. Every program aimed in one direction: toward a continuation of raising preferred crops on the same land. The agribusiness operators still choose to fight the environment rather than raise less vulnerable produce.

In the Northeast, from Massachusetts to Nova Scotia, a similar situation attracts attention each July. Enterprising landholders who have set out row after row of highbush blueberries become irate when more birds than people appreciate the genetically superior fruit. American robins, along with some catbirds, orioles, and other fliers, descend on the crop as soon as its color changes from green to red, to luscious blue. Enchantment with these native birds turns to rage.

Canadian wildlife officers recently killed thirty thousand robins in the Maritime Provinces to protect the cultivated

So long as an American robin catches earthworms and insects to feed its young and satisfy its own hunger, no one objects. But if these common birds of town and suburb eat more than a few cherries or blueberries that someone expects to harvest, human attitudes toward robins can quickly change. Even in Michigan and Wisconsin, where the robin has been chosen as state bird, it is expected never to compete for anything that enters commerce. (Photo by Jack Dermid, U.S. Fish and Wildlife Service)

blueberries. Within weeks after news of this slaughter appeared on television and in newspapers, letters of protest flooded the Canadian Wildlife Service. Indignation over the killing of robins far exceeded that over the killing of blackbirds and led promptly to a partial boycott of Canadian fruit. Yet no law had been broken. Not even the respected Migratory Bird Treaty between Canada and the United States had been violated. The outcry led to frantic tests of chemical repellents for future years.

One of our colleagues claims to have learned to think in three dimensions about blueberries and birds, not just in two. "It's quite a job," he admits. "But if I can fence out groundhogs from my vegetable garden, I can screen out robins from my blueberry bushes. The metal framework that supports my screens can stay in place all winter. The weight of snow and ice won't harm it. But before snow comes, I have to take down all the horizontal screening. I roll it up and stow it away. If I had to pay someone else to do that work each fall and spring, it would cost me most of the profit I get from the blueberries."

"Then what would you do?" we persisted.

"Raise something else, I guess."

The alternative—to destroy the competing birds—invites unforeseen consequences elsewhere. Canadian wildlife officers cannot offer a believable comparison in economic terms between the commercial fruit the robins eat and the farm produce they free from attack. Robins devour harmful caterpillars, beetles, bugs, flies, termites, millipedes, snails, and slugs. They also contribute to the welfare of many other forms of life by distributing seeds from the fruits they swallow. This adds diversity in vegetation, which is a tangible service to the wild community, even if it remains unmeasured.

Canadians should be equally concerned if fewer blackbirds, rather than more, fly north from the United States each spring. What changes due to burgeoning insects near the nest sites of blackbirds would ensue if few came to nest and pursue insects as food for nestlings? Grackles hold down the number of caterpillars in the spruce forests of Ontario and thereby protect the trees until they can be harvested for pulpwood. Blackbirds work unintentionally and unpaid for the newsprint industry. Redwings similarly suppress the insects in marshes of the North Country. This protects the plant cover that ducks and geese need while reproducing their various kinds.

We are mindful of the prairie pioneers in both countries. These brave men, strong women, and helpful children made the center of the continent safe for horses as draft animals and other domestic livestock. The pioneers freed the great area of large native life—bison, pronghorns, wolves, grizzly bears, and most of the birds of prey. The prairie falcon, which prefers prairie mice to domestic fowl, dwindled to the point of becoming an endangered species. The native prairie chickens mostly disappeared through a combination of intense hunting in open country and conversion of prairie into croplands, which support few non-migratory birds all winter.

The smaller life of the prairie continued for a while. Bobcats and coyotes hunted for prairie dogs and gophers, assuaging their hunger on mice of several kinds, rabbits, and even grasshoppers and other insects. Soon the ranchers began to calculate how many prairie dogs were equal to one cow or sheep in taking nourishment from green plants on western lands. That domestic livestock might eat plants different from those a prairie dog would choose hardly entered the mind of anyone. Now the "dogs," which actually are a sociable kind of ground squirrel, are almost gone. With them went a masked predator— the black-footed ferret—that had made the mistake of becoming a specialist, by living with, and on, the once-abundant prairie dogs. At present the ferret is close to extinction.

For years we praised the administrators of the University of Colorado for quietly protecting a small prairie-dog town on the campus. Naturalists among the faculty, students, and summer visitors parked by the hour along a narrow road to watch the short-tailed rodents. The prairie dogs soon lost interest in each vehicle as long as the people inside it made no obvious move. Out from the earthen doorways would come sharp-eyed sentinels and then young and old to feed on the short mixed vegetation of the unused field. Not one of the animals made a sound that we could hear unless a bicyclist rode past or someone approached a parked automobile. Shrill, sharp barks from any prairie dog would send all of them racing for home or the nearest hole. Just a few of the sentinels dared to remain in view, repeating the warning call until either the danger went away or came too close.

At least two hundred prairie dogs sustained themselves on that corner of campus adjacent to the state highway. Repeatedly their offspring tried to extend the town to the opposite side of the narrow paved road, toward the handsome

Prairie dogs, both white-tailed and black-tailed, no longer maintain almost continuous "towns" as underground burrow systems from southern Saskatchewan to northern Chihuahua in the American West. By every means possible, they have been killed, ostensibly to reduce the hazard to livestock, which might break a leg by stumbling into a burrow opening, and to increase the amount of forage the cattle and sheep can find. The change has nearly exterminated the black-footed prairie dog and has allowed sagebrush to spread over wide areas that formerly were much more productive. (Photo of prairie dog by D. A. Spencer, of ferret by Luther C. Goldman, U.S. Fish and Wildlife Service)

Kittredge complex of university buildings. Maintenance men held the line, grumbling to one another about the endless standoff, but sharing to some extent the delight of other people in seeing prairie dogs so close.

We returned to the Boulder campus of the university in July 1972, intending to record on film more antics of the prairie dogs. Audiences always enjoy the near-human way in which these little animals reach out with one paw to pull a flower head toward a nose and sniff it. The animals choose their food with obvious discrimination, sampling one item after another. Often they play with one another, although the oldsters at first resist, putting up a dignified front. Two dogs nuzzle briefly, press a paw against a chest as though to get attention for a joke, or appear engaged in friendly conversation. Always a few remain upright, alert for any danger.

That day an unfamiliar group of workmen hurried around the prairie-dog town. A conspicuous sign on their panel truck showed that they came from the Flatiron Paving Company. They unrolled long canvas hoses of the type used by fire-fighting companies and attached them to hydrants. From the truck they took out wire cages and strange tools like those used to capture poisonous snakes. Cameras in hand, we sought out the foreman and inquired about all this activity.

"We're going to move the rest of the prairie dogs," he explained. "Most of them were removed last December, but now we've got to finish the job. The Athletic Department needs this field for baseball practice. No one could use the field the way it is. You could break a leg in a prairie-dog hole. As soon as we get out all the animals, we'll bulldoze the area smooth."

"How do you propose to catch the animals?"

"That's easy! The fish-and-game department man was here yesterday and showed us how. We fill the burrows with water and help it penetrate by adding some of this." He held up a big plastic bottle of detergent. "The solution makes the prairie dog's eyes sting, and he hurries out of the hole. That's when we catch him with the noose on the end of the catcher stick. Into the cage he goes. When we've got all we can handle, we'll take them a few miles from here to another property owned by the university and turn them loose. The trip won't hurt them a bit."

We sped to the telephone and tried to get more answers from the university administration. Everyone with authority, it seemed, had left for the summer. We spoke to the zoologists on campus. They felt helpless, although the living treasure they had prized so much was to be taken away or left dead in the attempt. No longer would naturalists or visitors on campus be able to see this bit of the Old West at close range. Frustrated, we returned to the prairie-dog town to follow the action with our cameras. Why was no wildlife specialist supervising the work or even a representative of the Humane Society? Not one university official regarded this action as worth monitoring, to make sure it was managed properly.

The Boulder reservoir must have lost a lot of water that day. Thousands of gallons of drinkable liquid, not to mention dozens of bottles of detergent, went gurgling down the prairie-dog holes. Hour after hour the water gushed, as though to fill caves the size of Carlsbad Caverns. We wished for a Ralph Nader to add up the uncalculated cost of all this unmetered water, the hourly wages of the men who wielded the hoses but stood by, waiting idly to catch a prairie dog. Add the foreman's salary and the time out for several vehicles. How many prairie dogs got the message to emerge and be caught? We saw only four. One floated out, drowned, and was buried under a shovelful of earth. Three animals, miserably wet and shivering, went into a cage, where they continually rubbed their eyes.

By noon of the second day, the foreman decided that the job was done. He had treated every hole, including a considerable number that opened beyond the fence, along the highway right-of-way and not on university property. His men shoveled earth into the holes, more as a test to see if any prairie dog would still emerge than to begin grading the land for the Athletic Department. He avoided our questions as to the number of animals caught alive and transferred to the distant location. It may have been just those three. Or were they too killed to save a trip? The answer did not matter much, for the chance of survival for a prairie dog from one colony freed in the territory of another is extremely small. Nor would an animal half drowned and wetted with detergent put up much of a fight to live.

On August 1, the "Hotline" column of readers' questions in the *Boulder Camera* included an answer to "Whatever became of the prairie dogs removed . . . from the area behind Kittredge Common? Why were they removed? Why was one left behind?" The university sanitarian, Thomas LeMire, was quoted as favoring the move to eliminate a potential reservoir of bubonic plague and as promising to "look into the matter of the lonely prairie dog and see what can be done." Another university official, who declined to be identified, claimed the colonies of these animals were still thriving, undisturbed, on the university golf driving range (Potts Field) and around the university's underground cyclotron. The latter was the stated destination of any prairie dogs captured on the Boulder campus.

The high ground surrounding the university cyclotron, which was constructed in 1962, supported no prairie dogs until following the big flood of 1968. Then between fifty and a hundred prairie dogs arrived from adjacent inundated land and dug a town for themselves in the front lawn of the building. By 1975, only a few eroding burrows remained to indicate where the beleagured animals had been. Perhaps they were starved out of existence by a particularly dry spring season, when their fodder failed to grow normally. Or did someone fear plague or want a neater lawn? The Potts Field site appeared equally lacking in prairie dogs by September 1975. Apparently the university ball teams have inherited the only land of proven suitability for these engaging rodents. Yet the players use the area far fewer hours per year than the prairie dogs and other native animals did. Now, too, the service department must expend many man-hours to keep the field mowed. We can even mourn the wasted weeds that could have nourished a wide variety of different forms of life, sharing only energy from the sun.

Over much of the West, prairie dogs are disappearing rapidly through programs of poisoning, trapping, and paving. Animal behaviorists Jim and Judy Fitzgerald of the University of Northern Colorado, who seek to learn more about these social rodents, tell us that no sooner do they discover another colony to observe than someone comes along to exterminate the dogs.

"Who needs prairie dogs?" ranchers ask us.

In many areas we can respond honestly, "You need them!"

Prairie dogs turn over the soil, unsystematically but continuously. They convert vegetation into substances a carnivore can use. To prevent this utilization from seriously depleting the prairie-dog population, the alert rodents maintain visibility in all directions. This saves them from native hazards, both furry and feathery. Prairie dogs keep perennial sagebrush low and inconspicuous by chewing on the roots and eating new growth almost as fast as it appears. Without prairie dogs, the sagebrush grows three feet tall and becomes woody; it takes over the dry prairies, starving out the grasses and many of the herbaceous plants, such as lupine. It is lupines that, in flower, put purple in the "purple sage." It is lupines, grasses, and other plants, thriving where sagebrush does not shade them, that feed cattle and sheep, putting money in ranchers' bank accounts. Sagebrush itself is ash gray and fragrant, but a dominant pest—a blight upon land that was productive while it had help from prairie dogs. Cattle disdain sagebrush. Sheep will eat it while its buds and leaves are young but need other foods in sagebrush country to round out the year.

The diversified community of life in unforested areas benefits from the maintenance services of small rodents all the way to the arctic tundra. North of prairie-dog territory, which extends into Canada no great distance beyond the western United States, ground squirrels that hibernate have a corresponding effect. They accomplish their tailoring of the vegetation during the three or four months of warm weather, as they feed voraciously and selectively. In the Far North, small lemmings fill the same role. Wherever the climate allows the thin soil to support a growth of low sedges and grasses, the lemmings burrow extensively. Their food habits lead them to trim the rootless lichens and mosses, thereby favoring the tundra plants with roots several inches long. Sunlight can thaw the soil and offer energy to be spent in part on a seasonal carpet of flowers. Animals of many kinds take advantage of opportunities among these plants. But without lemmings, the mosses and lichens soon spread. They produce a thick insulation over frozen ground, shielding it from summer warmth while offering food to few kinds of animals. This answer to the question What good are lemmings? came to light only recently, through research on the fragile tundra along the Alaskan pipeline road. Between the principal pass through the Brooks Range and Prudhoe Bay on the Arctic Ocean, the ever-active rodents do far more than just nourish the watchful predators. Although preyed upon by arctic foxes, snowy owls, jaegers of the gull family, and sometimes huge bears, the lemmings contribute disproportionately to the welfare of species that scarcely notice their existence.

Rather generally in each wild community, every plant-eater helps sustain the precarious balance between the various kinds of vegetation as they compete for moisture, space, and light. The balance teeters dangerously if the plant-eaters become either too numerous or scarce. Fluctuations will occur at intervals, chiefly through variations in the combined effects of harsh weather, disease, and predators, all of which keep the system self-correcting if they are left alone. Yet the predators make attractive scapegoats. Although few, they tend to be conspicuous. Often they grow large enough to be noticed and move quickly at times in plain view. Removing them rarely contributes toward lasting stability.

Many a rancher tires of looking for competitors to exterminate. Never, it seems, can these confrontations with native life be won and done. Now some of these landholders, still intent on making a profit from livestock on open range, have come to suspect that predators and well-adapted rodents battle on the side of mankind in the endless struggle for survival on the inland plains.

"We've spent millions of tax dollars and lots of private money too," said one ranch-owner we met in Montana. "We've paid for bounties and traps and ammunition. It hasn't worked. Out in this country, nature is too big and human managers too few to change much. It's time to give collaboration a try."

"Does that mean you'll shoot the next coyote you see?" we asked.

"Not unless it's attacking some healthy critter. Maybe only a few coyotes are doing all the damage. I read somewhere that one out of every seventeen thousand human inhabitants of the United States is a convicted murderer and that five times as many at least escape conviction. It could be the same with coyotes, and other things we've tried to get rid of as varmints."

Already this man knew of the tests made in his own state and elsewhere aimed at predators (particularly coyotes) that kill sheep by seizing each victim by its neck. In flocks that are vulnerable, a few sheep are fitted with special collars, each almost covered with small, tough plastic packets containing crystals of potassium cyanide. Any predator that attacks in the customary way tears open some packets and gets a mouthful of the quick-acting poison. In each flock where this technique has been tried, no more than one or two further losses to predators have been experienced in the same growing season. Often unharmed sheep are found with a few packets ripped open, the contents washed away by rain, and no harm suffered by any animal in the flock.

"It almost seems that coyotes pick the sheep with the collar, then never live to tell another coyote it was a mistake," the rancher continued.

He was discovering also that predators elsewhere in the world are being reevaluated. Many emerge with good report cards and suddenly merit protection. In tropical waters of the eastern Pacific Ocean, for example, the spotted porpoises have earned a reprieve. These 250-pound fish-eating marine mammals accompany almost two-thirds of the schools of yellowfin tuna where fishermen search close to South America. As much as 80 percent of the tuna taken by boats out of San Diego are caught by locating the conspicuous porpoises, racing around them in a fast boat to lay a purse seine, then closing the net below the fish and porpoises at the same time. At this point, the porpoises become expendable. Are they competitors to be shot? Or are they merely unwanted because they may tear a net while attempting to escape? Or a nuisance when they get entangled and drown, unable to reach the surface for air? Why worry about unsalable porpoises, even if more than one hundred thousand of them are killed annually through the activities of tuna fishermen?

The porpoises may be neither feeding on tuna nor competing significantly with tuna for smaller fishes. No one has actually taken the time to learn what spotted porpoises eat. By contrast, the diet of yellowfin tuna is well known. It includes members of forty-four fish families and twelve different types of invertebrate animals, among them crabs and squids. Even if the porpoise's choice of food overlaps the tuna's somewhat, the Pacific waters may contain enough for all to share. Porpoises may actually herd the small fishes that tuna eat, thereby helping the tuna as well as making them easier to net. (Russian fishermen elsewhere report success by imitating the underwater sound of porpoises, driving schools of anchovettas and mackerel into gigantic seines.) And porpoises may protect tuna by driving away sharks. Any tuna a fisherman hauls into his boat from a school accompanied by porpoises is likely to be whole, undamaged, and profitable. But why invest in new nylon nets with finer mesh to reduce the chance of entangling porpoises? Why leave gaps in the net, perhaps near the surface, as escape routes a porpoise might find by echo location if some commercial fishes might be lost there?

Regulations aimed at safeguarding spotted porpoises from 1976 on, imposed by the National Marine Fisheries Service, merely irritated American tuna hunters. Their lack of compliance led in May 1976 to a new ruling by Judge Charles R. Richey, banning the use of all nets after the end of the month. It required a return to the fishing methods of the early 1960s, when live bait on lines from poles was used and there was no interference with the porpoises. Fishermen and porpoises can coexist while harvesting protein from the sea. Peace is possible, even if less profitable in the short run.

Even the fearsome crocodile merits a revised reputation. Until just a few years ago, no one possessed facts upon which to judge whether this ancient type of reptile was mostly detrimental or beneficial to the community of life. Everyone feared the gigantic jaws and lashing tail, because a crocodile can and will confront a full-grown person, often to the reptile's gain. This allowed Rudyard Kipling easily to distract us in his *Just So Stories* from any real answer to the question posed by the inquisitive Elephant Child: "What does the crocodile have for dinner?"

Now the Nile crocodile is extinct in Egypt. Crocodilians elsewhere are on the register of species in danger of disappearing altogether. Yet research reveals real benefits from their presence. Their dining habits change with growth.

Despite protection, American crocodiles and American alligators are not resurging in population in southern Florida, so visitors and residents no longer see as many of these powerful reptiles basking in the sun. One reason seems to be the release of young Central American caimans as unwanted pets that can compete successfully with the native crocodilians, without aiding other species in the Everglades by maintaining "alligator holes." (Photos of crocodiles and young caiman by Lorus J. and Margery Milne, of alligator by Luther C. Goldman, U.S. Fish and Wildlife Service)

236

When newly hatched, the little crocodile catches and swallows only large insects. Later and somewhat larger, the reptile takes small fishes and crabs, often those that otherwise feast on the eggs and young of commercial fishes. At the stage and dimensions that permit a crocodile to seize and eat waterfowl and mammals near the water's edge, the rate of growth has slowed and the number of meals required per month is much less. During all these changes, the crocodiles keep the aquatic realm open. They ensure a place for important fishes and lesser creatures, all without dragging any prey species toward even local extinction.

The American crocodile barely maintains a precarious place in southern Florida. It lurks around saltwater lagoons and inlets where boats go infrequently. For a number of years, a few of these reptiles caught rats where refuse from the town of Tavernier on Key Largo was used to fill in a mangrove swamp. Several naturalists and one photographer managed to see these crocodiles at close range. We spent hours one week trying to catch a glimpse in daylight, to no avail. Now the dump is closed, and further filling of the swamp forbidden. The rats have gone, and with them the attraction of the area for crocodiles. Only between ten and twenty breeding females of this species are believed to survive in the United States. They are too few to play any significant role in their environment.

For a while, everyone who knew about crocodiles in Florida assumed that they would benefit from the federal ban on commerce in all crocodilians native to the United States. This measure, aimed at both trade in alligator hides and the sale of young alligators as living souvenirs, should have allowed all of these reptiles to restore their former number in suitable areas. Alligators are appreciated most in Everglades National Park, because they maintain deep pools ("alligator holes") several yards across. These oases offer refuge during dry periods for fishes and for the big snails upon which the rare Everglades kite feeds almost exclusively. Herons, egrets, ibis, and other fish-eating birds retain a place in the community because the alligators are there. Knowledge of these relationships has led to reintroducing alligators in areas where they have been hunted to extinction. Louisiana has freed a sizable number from Florida in coastal swamps and bayous and enforces a prohibition against disturbing the alligators.

Some enterprising citizens expect alligator numbers to increase until legal harvesting of big ones for their hides can be permitted nationwide under strict control. Toward that day, these citizens have stocked alligator farms with eggs hatched in captivity. The young alligators are penned only with others of the same size and kept well fed, making cannibalism rare and growth extremely rapid. The principal investments are in wire for pens that are flooded shallowly, grain to feed the chickens that become alligator food, and dedicated labor to guard against losses from any cause.

Now wild alligators face a new menace. Ever since sale of young alligators was banned, souvenir sellers have imported caimans from tropical America and offered them as legal substitutes. Like any hatchling crocodilian, a caiman thrives in captivity and soon outgrows its welcome as a house pet. Thousands of them have been freed in Florida waters, discarded before they even left the state in which they were purchased. Caimans seem to survive so long as they escape frost. They prove able to best an alligator of equal dimensions and to kill and eat it. Caimans eat alligator eggs as well. Yet this progressive exchange of caiman for alligator represents a major loss, because a caiman maintains no equivalent of an alligator hole. If, as is possible, the substitute reptiles replace the alligators wherever frost is rare, the Everglades seem especially sure to suffer. The fragile community could degenerate into an unproductive sea of grass without its guardian alligators. Release of more alligators, obtained from alligator farms, may soon become an important service of the custodians of the national park.

As we follow these changing situations and the economic expediencies that lead so soon to new challenges, new tests of ingenuity, we realize the futility of trying to cope on a piecemeal basis. The ecological ship can scarcely be kept from sinking if leaks appear unpredictably in many places. Far better would be a collaborative program to get wildlife on the side of mankind. Exploitative operations could be changed into patterns that are totally self-supporting. Most essential is wide recognition of the reasons for change in policy. Explanation may be easiest in countries where literacy is general, even though there large segments of the human population live in cities, remote from the areas where wildlife is to be encouraged.

How, for example, should the government of the federal district in Mexico satisfy the three hundred day laborers who are newly redeployed, after working for many months along the canals of Xochimilco? These men are accustomed to clearing the canals for flower-bedecked flat-bottomed boats. Now the endless task is to be tended to mostly at night, by four full-grown sea cows—two males and two females—that have been liberated in these waters. Each sea cow is expected to devour daily approximately its own weight (550 pounds) in the aquatic vegetation that recently has clogged these waterways.

Manuel Cabrera Valtierra, from the Federal District Zoo Department, supervised the introduction of the sea cows where none had ever lived before. Privately he admitted to us that these lowland coastal animals would probably toler-

ate the colder water of the high plateau but not much mo-
lestation by residents or visitors who take too great an interest in their presence. Publicly he assured the villagers of
Xochimilco that sea cows are harmless and quiet. By day
they would likely stay at the bottom of the canals, avoiding
the many boats that are sculled or rowed along. Families
with picnic baskets, mariachi bands, and vendors with
handicrafts, flowers, fruit, or vegetables might never see a
sea cow from their boats. Only every twenty minutes or so
would the big animals need to rise to the surface for air.
Even then the blunt, bristly snout would appear only for a
moment and then disappear without attracting attention.
The timidity of sea cows may be their best protection.
Their instinctive life-style seems to fit them for inconspicuous collaboration with mankind and not for any confrontation.

Perhaps at Xochimilco, where the ancient Aztecs tended
their floating gardens to supply food for residents of Mexico City, the planned partnership between ponderous beast
and quixotic man can be sustained for the prolonged benefit of both. Yet we marvel that ingenious members of our
species have established so few of these mutually advantageous relationships. Virtually all of the domesticated animals were chosen for their roles a dozen millennia ago, before anyone had an overview of the world's wildlife and
plants. Progress with mechanical devices has turned fantasy
to fact. But possible avenues toward advantageous cooperation with nonhuman life remain largely unexplored.

How should those entrusted with responsibility for policy
decisions respond to fresh ideas and encouraging news?
The British government and private citizens alike could
find both cheer and motivation to make further improve-
ments in the river Thames, now that its waters have been
transformed through a dozen years of better care from a
grossly polluted and lifeless state to a haven for thousands
of birds. Mallard and pintail ducks have become especially
numerous, dabbling for food with teal, shelduck, and

widgeon. Pochard and tufted ducks dive to feed at greater depths or pursue prospective mates close to the reedy margins, where they find secluded sites for nests. Should bushes be encouraged along the gradually sloping shores, guarding against erosive destruction of steeper banks while providing more breeding territory for shelduck? Minor man-made changes could improve the habitat for wildlife, sponsoring still more diversity despite the nearness of the human population.

We ask the same kind of questions as we look at the Hudson River from New York City. Formerly this deep estuary offered valuable fishes, including female sturgeon as much as twelve feet long, to fishermen. As recently as 1911, John T. Nichols of the American Museum of Natural History could list 237 species of fishes found within fifty miles of central Manhattan. He noted that in one year, "domestic caviar" from local sturgeon sold for $6 million in the city markets. But something ended the resource. No one could be sure whether overfishing or pollution was the greater culprit.

In 1975, an adult female sturgeon was discovered in the Hudson. She was full of unlaid eggs, and the first record of this endangered species in the region for many years. Was the venerable, long-lived creature a sign of progress? Might other fishes be coming back into municipal waters? The sturgeon is a most deliberate feeder. Sculling with its up-turned tail, armored in its heavy platelike scales, it cruises slowly just above the bottom mud. Fleshy whiskers from the pointed snout just in front of the mouth serve like mine de-tectors. With them the fish identifies the position of each snail, clam, shrimp, worm, or other small slow-moving ani-mal. Alerted, the sturgeon protrudes its sucking mouth and siphons up the prey. Either living things of edible kinds or the flavor of fresh water at the Hudson lured the big fish into the broad channel. What more should now be done to improve the river, to make it attractive and safe for sturgeon and other fishes that could feed and spawn in the

A 31.5-pound sturgeon caught in Fort Peck Lake, Montana, is only about a third of its potential weight. The reappearance of this slow-growing, old-style fish in estuaries and freshwater lakes can be regarded as a hopeful sign that the environment is improving through attention to pollution control. (Photo by B. M. Hazeltine, U.S. Fish and Wildlife Service)

upper reaches? Any collaborative effort would affect a wide variety of fishes, for wherever a sturgeon can go to reproduce, other types of life would flourish too.

We feel especially concerned about other types of life.

Although the large and conspicuous animals and plants may serve to draw attention to changes in the environment, only occasionally are they the ones that contribute markedly to the perpetuation or the downfall of lesser kinds. Of course there are exceptions, among them, the American bison herds that kept the Great Plains grasses grazed and trampled, to the benefit of prairie dogs, black-footed ferrets, wolves, and Plains Indians. Or the giant redwoods, which condense showers of dewdrops from coastal fogs and thereby support a whole community of lesser plants and animals on ground that would otherwise be arid most of the year. But even more examples come to mind of inconspicuous creatures that wield great influence in their world. We suspect that they modify disproportionately the course of evolution, often without anyone noticing what is going on.

Long after an event, it may be possible to list the consequences. Harvard University's eminent bacteriologist Hans Zinsser offered his own compilation while arguing that the microscopic agents of cholera and typhus fever had influenced human history more than all the kings and military men combined. He mentioned some conspicuous contributions of the Old World to the New, such as Christian culture, trousers, horses and donkeys, Negro slaves, rum, and gunpowder. Along with these went inconspicuous mice, rats, and their lice; typhus fever; smallpox; measles; influenza; and tuberculosis. "For all these blessings it [the Old World] received in return at first only gold, tobacco, syphilis, potatoes, and Indian corn."

Zinsser did not mention that the slaves from West Africa brought blood flukes to the New World. The debilitating infection became endemic to parts of Puerto Rico, the island of Saint Lucia in the West Indies, and parts of South America. Until 1973, parasitologists could not understand why African blood flukes did not gain a hold also in Saint Vincent, which is the next island south of Saint Lucia, or in Grenada or Trinidad, still farther to the south. Then someone noticed that the famous guppy of northern South America ranges as far north along this enchanting archipelago as Saint Vincent. Guppies, never more than 1.5 inches long in the wild, pursue and destroy the infective swimming stage of the introduced blood fluke. They and small fishes with similar habits elsewhere in the New World may well have determined how suitable each environment would be for reinfection of still-ignorant mankind. Who would have credited a guppy with being able to do anything important, unless perhaps to demonstrate live birth in a home aquarium? Now we can list guppies as protectors of mankind!

Any collaboration requires that both principal beneficiaries make a contribution. Support must be available in each direction, to whichever participant needs it most at any given time. Often human activities affect many species simultaneously, while the effects in the opposite direction may be evident from only one or two species. Even then, the underlying explanation frequently eludes notice, because the causative agent is so inconspicuously small. We can hope that scientific probing will lead to discovery of the true situation before some unforeseen extension expands the adverse effects rather than the beneficial ones and puts the ecology out of joint.

An instance of this kind turned up recently in our fiftieth state, Hawaii, where nearly two-thirds of the islands' most distinctive birds—the honeycreepers—became extinct soon after Christian missionaries began introducing European ways. Of the twenty-two species, many with several distinctive races on different islands, some were extremely colorful and used their curved beaks to probe into flowers for nectar and small insects. Other honeycreepers, with shorter and straighter beaks, combined insects and seeds in their regular diet; these birds, for the most part, wore inconspicuous green feather coverings. Yet the extinctions occurred chiefly among the second group of honeycreepers, not those whose bright plumage appealed to the Polynesian settlers of Hawaii. No one could blame the pagan people for having exterminated these native birds to obtain feathers for ceremonial robes, although great numbers of brilliant honeycreepers must have been killed for this purpose. Nor did the major losses among the species and races of the peculiar birds coincide with the twentieth century, when Hawaii, first as an annexed territory and then a state, underwent great changes in land use to accommodate the sugarcane, pineapple, and tourist industries.

Ornithologists regarded the disappearance of the honeycreepers as more mysterious than their origin and diversity. Why should so many different types have been able to survive with Polynesian neighbors from about the fourth century A.D. through the eighteenth, only to die out in the nineteenth? It was easier to account for the presence of these intriguing birds by assuming that a pair of ancestors, probably buntinglike migrants from tropical America, managed to cross the two thousand miles of open Pacific Ocean to Hawaii about five million years ago. That was soon after the remote volcanic islands acquired their first forests and could accommodate arboreal birds. Thereafter the ancestral honeycreepers could have diversified to take advantage of types of food and nest sites that had been unavailable to them on the American mainland. Those descendants that evolved suitable beaks and bright colors could become pollinators of Hawaiian plants, frequenting their flowers in ways roughly comparable to New World hummingbirds, African sunbirds, and New Zealand flowerpeckers.

The true cause of decline of Hawaii's honeycreepers has now been identified: infected mosquitoes. No mosquito of any kind was known on the islands until 1826. Then a ship from Mexico docked at Maui, and from a barrel of water on the deck, a few mosquitoes flew ashore. The particular species and geographic race of this new addition is shown by its scientific name (*Culex pipiens fatigans*). The insect can develop in brackish water almost a third as salty as the ocean or in fresh, can fly fourteen miles, but not tolerate the cool weather of Hawaii above elevations of three thousand feet. Through all the lower areas on Maui, and later on other islands as the insects rode passively on shipboard from place to place, the mosquitoes bit both people and birds. They carried no disease to mankind but transmitted bird malaria and bird pox. These infections proved fatal to most honeycreepers that nested below three thousand feet. The surviving kinds are those that use sanctuaries higher on the mountain slopes.

In North America, the Mexican mosquito has a counterpart known as *C. pipiens pipiens*. Only a specialist can tell it from the race in Latin America. But the more northerly one prefers fresh water for its eggs and immature stages and tolerates cold quite well. Entomologists at Cornell University in New York State have caught this species flying every month of the year, even on warm days in midwinter. It is a nuisance mosquito in northern Canada and Alaska each summer. And if it can get to Hawaii, it will almost certainly spread bird malaria and bird pox to the remaining kinds of honeycreepers, ending this evolutionary line in short order. Without the colorful species and races that visit flowers and carry their pollen, many of the flowering plants that presently reproduce well at higher elevations in Hawaii would be in trouble. They have become dependent on the services of these particular birds and might soon die out. No longer can we doubt that indirect effects of barely noticed change can be far-reaching. Diversity throughout the world needs all the care we can give it, even in minute details.

How much more comfortable our human stance would be if we scrupulously avoided redistributing life on earth. Mankind is an endangered and a dangerous species only through grasping attitudes which lead to confrontation. Instead, for the good of all, we might try in every imaginable way to go forward side by side with other kinds of life. We share the same recycling elements of earth, air and salty ocean. But it is nonhuman life, not we, that can renew this tangible resource. We share the same energy from the sun and get ourselves in trouble whenever we attempt to mine for extra energy in other ways.

"We can never have enough of nature," wrote Thoreau. "We must be refreshed by the sight of inexhaustible vigor, vast and titanic features, the sea-coast with its wracks, the wilderness with its living and decaying trees, the thunder cloud, and the rain which lasts three weeks and produces freshets. We need to witness our own limits transgressed, and some life pasturing freely where we never wander."

We need the small things too, filling the fine mesh of ecological relationships. The future of life on earth depends upon small specifics and not on approximations, upon details and not generalities. It is the particular tree rather than the great forest or the individual grass plant rather than the prairie that carries the genetic heritage concealed within. That heritage is a coded program for living in a particular way, one that for millennia has brought success. The coded details have been located within specific molecules. But although all follow the same code, they are so numerous that no team of scientists can decipher them meaningfully. A small bacterium has about twenty thousand specifications, a person, nearly eighty million. They determine the pattern for each kind of life. Who has the knowledge to decide which species have a right to perpetuate themselves and which can be obliterated? Lacking such nearly divine wisdom, mankind might better safeguard by every possible means the genetic diversity that remains.

10

Unequal Horizons

ONE OF OUR CLOSE FRIENDS feels no reluctance to change his plans from day to day. "Look," he insists, "I'm not the same person I was yesterday. I've grown. I've learned new facts. Why should I persist in a choice made on the basis of incomplete information? My horizon is farther out. My perspective is different. That makes me different too." He strives to stay flexible, the better to meet each tomorrow.

Situations do change and justify fresh planning, often more than anyone notices for a while. Census information, summarized each decade, reveals not only the overall increase in human population but also shifts in where people live. These facts, in turn, reflect alterations in the use of land. Frequently they make obsolete the reasons for which large animals were banished years ago from some parts of America. They also render suspect some continuing laws, enacted to protect certain kinds of life that now have burgeoned to embarrassing abundance.

A number of families in North America today enjoy tracing the changes that affect a particular piece of land over several generations. One of us (L.J.M.) notes the receding frontier from a site on Yonge Street, a central north-south highway through metropolitan Toronto. He was born there, in a brick house built in 1801 by his great-great-grandfather. The land on which the structure still stands had been conveyed on September 4, 1800, by King George III as a crown grant to "Abraham Johnson, Yeoman." We treasure the parchment deed with its dangling 4.5-inch, 5-ounce royal seal. We cherish also the pelt and tail of a gray wolf that Grandfather Johnson shot on the modest farm that his grandfather had hewed from the wilderness. At that time, Yonge Street along the eastern boundary of the farm was a rutted wagon road. It extended nine miles south to the docks on Lake Ontario and just slightly less to markets where produce might be sold.

Grandfather Johnson did not hang the wolf's body on a fence to warn off other predators. Instead, he tanned its hide. Grandmother Johnson sewed the pelt to a rectangle of heavy cloth to make a "wolf blanket." It kept her infant

The wolf is well established in wild parts of Canada but remains a controversial predator in Alaska and south of the great Dominion. Even on the sanctuary of Isle Royale in Lake Superior, where wolves have maintained a fluctuating balance with moose and beaver, the maturing of the forest is reducing opportunities for all three species. (Photo by C. J. Bayer, U.S. Fish and Wildlife Service)

daughter warm in winter as a covering in the baby carriage. In due course, it held in the heat for her grandson too. A photograph records the outdoor juxtaposition of the wolf blanket, the grandson, and Grandfather Johnson. The blanket remained at other times in mothballs as an heirloom.

Today, metropolitan Toronto extends to within two miles of the family farm, and a subway under Yonge Street speeds people north and south. The brick house was expanded to become an orphanage (York Cottage), while the farm's 210 acres were built upon and paved over as a subdivision, complete with shopping mall. The only wolves closer than the lake front now live in the city zoo. Yet their relatives roam wild fewer than a hundred miles to the northeast, in Algonquin Provincial Park. This sample of the old frontier was set aside in 1893, to preserve 2,741 square miles of wilderness and wildlife in northern forest.

Any wolf is safe to howl and socialize or pursue natural prey in Algonquin Park. It is there that scientist-photographers for the National Parks Board of Canada record on color film and magnetic tape the behavior and sounds of these native predators. One documentary bears the provocative title *Death of a Legend*. Does it mean that wolves are doomed? Or that the legend of their ferocity rests on folklore rather than fact? Is our modern horizon farther out, our perspective different, and the wolf worth saving after all?

Everyone, it seems, encounters at an impressionable age the famous painting of a desperate Russian family in an old-style sleigh, pulled by a troika of galloping horses, followed by a pack of ravening wolves. What next must be jettisoned to the fierce beasts to slow them down and let the family escape the fearsome jaws? Yet when Canadian wildlife biologist C. H. Douglas Clarke recently investigated the many claims of wolves attacking people in the whole of Russia, elsewhere in Eurasia, and in North America, every record fit some one of three categories. Most numerous were those for which specific details are lacking. Second are instances of individual wolves with rabies, biting and slashing in a pattern of behavior utterly unlike any normal activity in seeking prey. Finally are two extraordinary wolves that killed almost a hundred people in the old French province of Languedoc. After the two beasts were shot and examined, they appeared to be rare natural hybrids between wild wolf and domestic dog, combining special vigor with a reduced fear of mankind.

The wolf relies upon its keen senses to avoid people, as well as to interact with other members of its group and to detect members of the deer family, which comprise its principal prey. Only once has one of these wary predators been outwitted by the leading student of wolf behavior, L. David Mech, who was able to photograph the wild animal from a distance of fifteen feet. Mech and his bush pilot in a small plane had been following a pack of wolves above the frozen shores of Isle Royale, a small forested island of Michigan in Lake Superior. The wolves set out in single file across the ice of a cove. Ahead of them, on the farther side, stood an abandoned fish house. Mech had his pilot set down the plane on its landing skis out of sight of the wolves so that the two men could hurry to the fish house and hide inside its partially open door. A big male came closer than expected. He cocked his head at the click of the camera shutter but continued on. Mech regards that moment as one of the triumphs of his life. It was a high point in his study of the wolves of Isle Royale, which earned him his doctorate degree at Purdue University.

No wolves lived on Isle Royale prior to the particularly cold winter of 1949. Then a pack arrived from Canada across the frozen lake. They found a population of moose that had been burgeoning and crashing with monotonous regularity since at least the turn of the century. That was when new forest growth following logging operations offered abundant food and moose found it. The big browsers multiplied unchecked without wolves, until as many as three thousand fed from the vegetation, which could not sustain them through the snowy months. At about ten-year intervals, most of the moose herd starved to death in late winter, amid the devastated vegetation. A few moose survived to start the cycle anew while the shrubby growth regenerated. Tree rings reveal the history.

Wolves on Isle Royale slow the increase in moose numbers by eating young moose that stray from their mothers. The wily predators kill other moose that are sick or injured. Unwittingly the wolves help hasten this turnover in nourishment by distributing in their droppings the infective embryos of a hydatid tapeworm. The young worms, surviving as much as a year in mere dust particles, develop in any moose that chances to swallow them. They interfere with the functioning of its internal organs before settling down, ready for the next move when devoured by a wolf. The cycling of the parasites between wolf and moose, the effects of other ills to which moose are prone, and the activities of vigorous wolves combine in holding the moose herd on Isle Royale to about six hundred animals. The forest grows unharmed. Most moose seem healthy, well able to fend off a pack of predators.

The number of wolves remains stable too, at fewer than thirty. The dominant male and his consort in the larger pack eat well. They succeed in bearing and raising young each year. The smaller pack has less success in hunting and gets less energy to share in reproduction. The few outcasts and loners live precariously and contribute nothing to posterity. They do, however, affect other native animals. Particularly in spring, before the pregnant moose have given birth, the wolves dine on beavers. At that season the beavers have just emerged from a long winter below the ice of the beaver pond and in the beaver lodge. Their store of branches with nutritious bark is almost gone. The beavers take chances to reach fresh food, and the wolves benefit. The predators get young beavers too, individuals that have just been evicted from the lodge to fend for themselves while their mother gives birth to a new litter. In some years, beaver meat accounts for fully 10 percent of a wolf's annual diet. This rounds off the crests in the rise and fall of beaver numbers on Isle Royale and prevents the big rodents from destroying too many trees.

Like other scientists who have made a pilgrimage to Isle Royale to confirm for themselves the balance linking the vegetation, the plant-eaters, and the big predators, Robert Forster of the Canadian Forest Service appreciates the strong web of food relations. "But already," he tells us, "the number of moose, beavers and wolves is showing a perceptible decline. Spruce and fir are replacing the forest of mixed deciduous trees. They will create a dense stand of just the type favoured by the island's northern climate. It's almost a bit of Canada, you know, just thirty miles from the metropolis of Thunder Bay!" And, like other foresters, he offers two suggestions to improve the environment for moose and wolves: either a logging operation in clear-cut blocks or a program of controlled burning to set back the woody growth to a more productive stage. Either way would increase the food for moose and beaver and sustain more wolves too.

Wildlife biologists elsewhere in Michigan still admire this example of unmanaged interaction between wolves and their natural prey. Other wilderness areas in the northern part of the state could support wolves too. To convert these judgments into action, a cooperative effort began in March 1974 to introduce a small pack from Minnesota, which is the only state south of Canada with free-roaming wolves on a mainland. The fish-and-game commissions of Minnesota and Michigan, the U.S. Fish and Wildlife Service, the National Audubon Society, the Huron Mountain Club, and biologists at Northern Michigan University arranged the details. A dominant male wolf, his consort, and two younger animals (a male and a female) from the same pack were trapped. Equipped with radio transmitters on harmless collars, the animals were freed in the wild and roadless Huron Mountains on the southern shore of Lake Superior.

Promptly the dominant pair and one male wolf headed west, leaving the young female behind. Later the three turned southeast and appeared to establish a home range for themselves in Iron County. There the pack leader died, struck and killed by a car in mid-July. The second male died of small-caliber gunshot wounds late that month. Before the end of the year, all four transplants were dead, every one of them through an encounter with mankind.

Twice we have had our caressing hands on a live gray wolf that had not yet learned to distrust people. Each time we recalled the words of Shakespeare's wise Fool in *King Lear*, "He's mad that trusts in the tameness of a wolf, a horse's health, a boy's love, or a whore's oath" (act II, scene 6). Each time, the animal tolerated our touch but essentially ignored us. Its eyes, its sharp ears, its magnificently sensitive nose were tuned for other cues, for signals we could not detect. Probably it sought reassurance from other members of its pack, for a solitary wolf can rarely relax. Only once did we see one relish its solitude, to roll ecstatically on clumps of fragrant tansy. Then the animal sprang to its feet, ready to leap away. Its horizon clearly exceeded ours, particularly when we knelt to bring our eyes and ears to the same distance above the ground.

Probably it is inevitable that people with a charitable attitude toward wolves will suggest their reintroduction in many parts of the country to enliven the wilderness scene. Five organizations that promote conservation causes banded together in such a request to the governor of New Hampshire in 1976. They urged the restoration of wolves in the White Mountains area of the state, where the human population is low and public access quite limited. Supposedly so large a tract of land could support a population of the predators without endangering human life or domestic livestock, and hence "minimize potential conflicts with human economic interests." But the quiet forests in which introduced wolves would be expected to establish home territories grow denser, with fewer openings, less low vegetation, and fewer deer every year. They afford almost no moose, elk, or other wild meat in sufficient quantity to support a wolf pack, to keep the predators from going elsewhere in search for a better living. Nor would the scattered landholders nearby, who try to make ends meet on small farms, welcome a visiting wolf pack. Although uneconomic anachronisms, these personal enterprises still receive far more sentimental support than any program to restore the wilderness wolves. If a wolf killed a farmer's cow, no one would judge the loss in terms of the degree of care the owner gave the animal, its age or condition, or the importance of his continued operation on the land as shown by taxable income. The wolf, not the independent farmer, would be the one indicted as out of place.

Any proposed reintroduction of a large predator should be based upon a realistic measure of its probable success, related to the available food supply rather than mere land area. "Live free or die" is the state motto of New Hampshire. Yet neither a person nor a wild creature is likely to be limited to these alternatives if actually hungry. So long as legs can provide transportation, the individual can move. Elsewhere the opportunities may be more favorable. Indeed, the present areas where wolves might be released include many an abandoned farm, the owners of which moved on. If the climate favors the colonization of the land by forest trees and neither fire nor forester interferes, the vegetation progresses through regular stages of succession from weeds to trees. The foliage upon which plant life depends in capturing solar energy rises higher above the ground. Soon it is out of reach of rabbits and deer, and the region no longer can support so many animals that walk upon the earth. The predators have to travel elsewhere, even far away.

Do members of our species choose the wolf as a scapegoat to avoid admitting environmental change or merely envy the controlled power of this magnificent animal? In Minnesota the commissioner of natural resources asks that the wolf be removed from the list of endangered (and hence totally protected) species. He seeks a legal hunting season for wolves, in part to satisfy those who claim that wolves are responsible for the decreasing number of deer in his state and in part to end occasional wolf attacks on livestock where wolves find few deer to eat. He must be aware that as few as 350 wolves survive in Minnesota, almost all of them protecting the Superior National Forest from unhealthy browsers—deer and moose. That forest has been maturing since early in the century, when lumbering was prohibited. So long as it is kept unmanaged, it will provide progressively less browse and nourish fewer deer, moose, and even wolves. This particular wilderness along the Canadian border is barely bigger than Algonquin Provincial Park. Yet President Truman in 1951 recognized its uniqueness by banning all aircraft flights over it at elevations of less than four thousand feet. The wolves cannot be blamed for that decision, or the earlier ones either.

In Alaska, we realize, citizens who call themselves sportsmen seek to legitimize the hunting of wolves with aircraft and high-powered rifles with telescopic sights. No one knows how many or few wolves remain in this vast state either. The number may be as small as seven hundred. They harm no one and help suppress periodic changes in the abundance of prey animals on a challenging landscape. The Scandinavian countries have between them only about thirty wolves, and Japan, none at all since the end of the last century. The remaining wolves are still hunted avidly in China, the Soviet Union, eastern Europe, the Italian mountains, and the few patchy wilderness areas of Spain. Canada now holds the key to the future for this oncewidespread and efficient predator. How the wolf fares will be a test for all humanity.

Wolves were common throughout North America in 1636, when John Harvard founded in Massachusetts the college that bears his name. So were the big native cats that New Englanders know as catamounts ("cats of the mountain"), mountain lions, pumas, or cougars. A colonist could almost count on four wolves and one cougar in each sixty square miles of good hunting territory. This area, about nine miles across, could be patrolled by a solitary cougar. The wolves came and went. Between them, these predators kept the prolific deer and elk from outstripping the woodland food supply.

Until Harvard University celebrated its first century, cougars were still the most widespread members of the cat family on earth, ranging from Atlantic to Pacific coasts, from Alaska to Tierra del Fuego. But many of them learned how much easier life could be if they killed a horse or a cow every other night, instead of a deer or an elk. This adjustment saved the predator much energy but brought it into immediate conflict with the human population. No one philosophized that this change in habit represented a tradeoff. The population of deer and elk could increase when neglected by cougars, giving the colonists more targets in hunting season. The cost of livestock was too great. Cougars had to go. So did wolves, which found domestic sheep a fine substitute for deer and moose.

In retrospect today, we can admire the balance of nature that the cougars and wolves maintained before firearms and steel traps were introduced. Each small-headed, long-tailed cat might weigh as much as 200 pounds. Characteristically it pounced from an overhanging tree limb onto an unwary deer or elk, killing its victim quickly. The cougar ate seven or eight pounds of the choicest meat and then tidied up the site by covering the rest of the carcass with leaves. Actually, the sated animal abandoned the carrion to bobcats, foxes, and other denizens of the woodland. A family of wolves, by contrast, ate most of the one big prey they killed each week. The 150-pound adult male, his 80-pound consort, and their two 75-pound youngsters ate as much as a fifth of their own weight at one sitting. If their prey was

Seldom seen, the cougar, or mountain lion, is a long-tailed native cat of American forest edges, formerly more numerous from Alaska to Tierra del Fuego. By preference, a cougar weighing as much as two hundred pounds will kill a deer every other night, eat a hearty meal, and leave the remains for scavengers. This behavior is an important contribution to preventing deer from increasing in numbers beyond what their food supply will support. (Photo courtesy of U.S. Fish and Wildlife Service)

large, each wolf would dig a hole, disgorge into it a first meal as a cache to be consumed later, and then return for an equally generous refill. Together the family could dispose of 150 pounds of meat and leave little to ravens or other carnivores. A wolf family could sustain itself on one adult deer or elk for a week and thereby draw down the prey population only about a third as much as a single cougar in the same hunting area.

The 60 square miles in a cougar's territory might well support 768 deer, each nourished by the greenery on 50 acres. Healthy reproduction would increase the herd by about a third (259) each year, unless the equivalent number were "removed." The cougar could account for 71 percent of this normal surplus, and the wolf, 20 percent. Nine percent would be a reasonable decrease due to accidental death, diseases, small predators, and human hunters. Starvation need claim no deer or elks in this balanced transformation of plant foods into meat for carnivores, and the forest could remain healthy.

Today, starvation and roaming packs of dogs substitute for the native predators. Licensed deer hunters—most with guns but a few with bow and arrows—take the place of the Indians. It is difficult to prove that, despite better equipment, they kill a larger percentage of the deer than the Indians did in earlier times. The number of deer is much greater, just as is the number shot. Accidental death and diseases claim more. Yet in many states, the deer have become a blight upon wild land.

The situation in Vermont seems particularly bleak, since that state now leads the nation in the scarcity of wild acres available to its white-tailed deer. Statistics suggest that a third of Vermont is forested, a third is farmland, and the rest bare rock, paved or built upon. This affords between ten and twenty acres apiece for the nearly two hundred thousand deer. In winter, they are still more concentrated in "deer yards," where they seek seclusion and food. During those cold months most does are pregnant and better fitted for survival than young animals or bucks because they still have a reserve of fat from summer feeding. The bucks in Vermont are undersized—about thirty pounds lighter and several inches smaller than their counterparts in adjacent New Hampshire. They begin the winter fatigued and malnourished from their activities during the autumn rut, with little fat to tide them over until spring. Young animals, particularly those born late to young does, are at a special disadvantage because their needs for food are greatest while their small size prevents them from reaching much of the available browse.

From 200,000 deer, we might expect more than 66,600 fawns each year. An equal number of mortalities would hold the population steady. In Vermont, about 46,000 deer starve to death during the cold months, or are destroyed by roaming packs of dogs. Fewer than 8,500 deer are taken legally in hunting season. Almost 2,200 collide fatally with motor vehicles on highways and back roads. This annual tally of 56,700 deaths overlooks animals that perish from other accidents, illegal hunters, small predators, and contagious diseases. If these were added, the total would still not reach that of annual births. A surplus of survivors continues to swell the deer population.

Ben Day, the chief game biologist for Vermont, worries because the starving deer now are moving onto farmland each winter. Already they have caused more than a million dollars' worth of damage to forests. "Everything that one day might become green is eaten," he says. "They chew the bark off trees for as high as they can reach."

Possibly the laws regarding deer are more antiquated in Vermont than in other states. Management of these animals has been a responsibility of the legislature since 1865. Each year, the Vermont Fish and Game Department goes through the ritual of asking the legislators to authorize the hunting of antlerless deer so that does may be taken as well as bucks. A limited season has been opened for both sexes just ten times in 110 years. Hunters enthuse less over shooting does than bucks, even if hunting season ends before they kill a deer at all. Consequently the legislators let old rules stand and try to ignore the complaints about deer damage by foresters and farmers and about the size of deer by the hunters who succeed in shooting a buck.

North American laws, originated by the colonists, make deer the sacred animals of the continent. These native wild browsers receive resolute protection despite the damage they cause to gardens in summer, orchards in winter, and

A white-tailed deer fawn, motionless and well camouflaged in the dappled shade near the edge of a mixed hardwood forest, may be one of two born to the same mother in quick succession. The twin would be hidden nearby, sought out and given a good meal of milk after dark each evening. Male deer have no part in protecting the young. (Photo of fawn by Marvin Lee, of buck by E. P. Haddon, U.S. Fish and Wildlife Service)

forests and motor vehicles at any time of year. Deer are special wards of the state, ranking higher than any taxpaying citizen. They are free to come and go, to eat whatever they fancy without reprisal, unless a landowner has erected a deer-proof fence at personal cost to keep them out. The estate upon which a deer is born, feeds, and dies gains no right or recompense. Indeed, the legal penalty may be less for shooting some farmer's cow than for shooting a deer out of season, without a license, after dark in season with a license, with an automatic weapon, with a gun on a tripod, or for selling the meat (venison) of a deer shot legally. Even a motorist who salvages the carcass of a deer that has killed itself and damaged his automobile by unexpectedly dashing across a highway is liable to prosecution if he does not go directly to the nearest game warden and surrender the body with full explanation. To leave it on the roadside is a hit-and-run offense. The cost of the delay and of repairing the vehicle falls on the owner, not the warden or the state. The deer retains the right-of-way, with a legal status scarcely lower than that of a human pedestrian.

Outdoors people in other countries ask us about the deer in America. "Could your native predators not be allowed to fulfill their inherited roles? Or should your state officers not cull the surplus animals in whatever way, season, or hour of day would be most efficient in reducing numbers?" We have to admit that licenses are still issued to allow the killing of wolves and cougars in a few states that have some of these powerful carnivores at large. Yet we know that most reasons for this policy have disappeared. The vulnerable horses have been replaced by tractors. In most of the country where deer are numerous, the cattle and sheep are safe in the barn each night and outdoors only in small fields by day. The attitude toward large predators that made sense three centuries ago is largely obsolete. But tradition continues to regard more deer as good and anything other than a human hunter as bad if it kills a deer. We feel relief that bounties are no longer paid to encourage hunting of those agents of natural control. Yet we also envy the New Zealanders who freed themselves from the old ways when hunters failed to cope with the natural rate of reproduction of game animals.

Perspective is different now over much of the world. The American mining engineer Rossiter Worthington Raymond can admit in a prayer for the future that "a horizon is nothing save the limit of our sight." The poet Theodore Roethke can enthuse over the long view, particularly on the open plains where "our feet are sometimes level with the sky." Now that the simple solutions and simple communities of life have proven unreliable and unstable, it seems time to abandon the old dream of conquering nature. Instead, the wild animals and plants that mesh together so advantageously could be taken into partnership with the human species, to the benefit of all.

Surely a team of ingenious planners could assemble the necessary information and build a self-sustaining system composed of living things, some one or more of which would be useful to mankind. The experts would have to calculate how much each component was needed, how much it would contribute to the total enterprise. In lieu of solar panels to capture energy from the sun, the most tested and refined energy-capturing mechanism in the world—green leaves—would serve. Living cells would store molecules of organic compounds from which the energy could be released later, as though from storage batteries. To provide for unforeseen situations, it would be well to have at least three different kinds of green plants, carefully selected, any one of which could temporarily sustain the system. Beech, black birch, and sugar maple appear to interact in this way in a mature New Hampshire forest.

In this living system, judiciously arranged to ensure an ongoing supply of human benefits, many other kinds of life would be needed to adjust the components and to recycle every waste. Each species offers its own self-replicating inventory of parts and capabilities, as though combining a miniaturized computer and complete production facility. Extra kinds of living things, equally well chosen, should be included as backup equipment, in case some link in the system shows deterioration. These extras would be redundant under predictable conditions. Yet their presence would, at little extra cost, ensure the stability of the whole functioning network.

In natural communities of life, the stabilizing systems are so effective that their role often goes unnoticed. Particularly in a tropical rain forest, nothing seems to change unless with cyclic, seasonal regularity. Myriad interactions damp out all fluctuations where the complexity of the food web is so overwhelming. Researchers feel more at home where the number of different kinds of plants and animals is much smaller, as in temperate regions. There fluctuations occur so unpredictably that their causes might be found. Even then, any close exploration reveals an amazing wealth of backup species, which keep the natural system from getting irretrievably out of joint.

James W. Tilden tells us what he finds on a shrub with clusters of thistlelike flowers on dry hillsides of California, close to his home in San Jose. Deer browse on this shrub, which is known either as chaparral broom or coyote brush. It supports more than its own welfare and that of deer but seems to require much the same from its environment as other neighboring plants: moisture from morning dew and occasional rain, carbon dioxide from the local air, small amounts of dissolved minerals from the soil. Chaparral broom attracts at least 257 different small animals with jointed legs (arthropods) directly or indirectly. Of these, 36 are spider mites and spiders, some of them sucking juice from the plant while the rest prey on insects. Of insects, Tilden recognizes 221 species, 53 of them sucking sap, nibbling plant tissue, eating pollen, or causing formation of galls. These 53 primary herbivores are eaten by 23 different kinds of predatory insects, as well as by spiders and spider mites. At least 65 types of parasites come in for a

share, 35 of them attacking specific herbivores or predators, 9 attacking other parasites, and 1 making a living at
the expense of a parasite on a parasite! All of these animals, and probably more that go unnoticed, gain their energy from the supply that represents a surplus captured by
the green plants. At the same time, the animals limit each
other's impact and thereby leave the chaparral broom
healthy and secure.

Even a single wild species may be critically important to
another, without the relationship becoming obvious. We
ourselves might never have guessed that the date of arrival
of a migrant bird made any real difference, had we not
kept records for more than a dozen years on two seasonal
events that entertain us. The two combine just below our
bedroom window, where a score of Solomon's seal plants
from a particularly large and handsome variety grow. In
ten years out of the twelve, the whirr of hummingbird
wings has awakened us before dawn on the first morning
when those Solomon's seals opened their paired pendant
bells to any pollinator. Each time, the date has coincided
with our first sighting of a hummingbird that year anywhere in town.

We joke with one another about the regular rendezvous
of the birds from their Central American wintering area
and the plants that arch upward each spring from the soil
below our window. Yet the two years out of twelve when
stormy weather in the states to the south along the Atlantic
coast held back the migrant hummers for about two weeks
have been the only ones when those Solomon's seals produced no fruit. The plants grew on schedule. They opened
their bells and offered nectar. But the flowers withered
before the long-beaked hummers arrived. No pollen was carried from bell to bell. We wonder how many other springtime plants failed to fruit for the same reason in those
two years. The next time the hummingbirds are late, we
will take special note of wild Solomon's seals to see whether
their lower growth and shorter flowers make them accessi-

ble to another pollinator. Perhaps a springtime moth with a long tongue is a natural associate, a backup that helps stabilize the reproduction of the plant. Our giant race may simply outgrow the community in which its ancestors maintained a secure place.

In the Temperate Zones, our experience with plants and animals that fluctuate in numbers from year to year makes us expect some versatility in associations. We know how many different flowers will attract a hummingbird or a moth, a butterfly or a bee. We realize that a rabbit will eat many kinds of plants, not just carrots and dandelion greens. Yet these reliable observations yield an inadequate perspective to appreciate the critical times in the life of each familiar creature. The moth or butterfly, although well nourished for a short adulthood by sips of nectar, may still have difficulty finding the particular plants upon which to deposit eggs. Here its choice is often narrow. Actually, the number of kinds of butterflies we see in a summer measures reliably the diversity of wild plants in the area. In the British Isles and Australia, where the number of European rabbits has been brought low by means of the introduced disease myxomatosis, wild flowers that had rarely been found for many years are now spreading. With them has come a resurgence in native butterflies and moths that had been scarce for a century or more.

Often we relegate to children and the unusual adult an intense interest in moths and butterflies, without realizing the significance of these insects as indicators of our environment. For every two kinds of seed plant on earth, there is one kind of butterfly or moth. Of all the orders of animals, only the beetles outnumber the butterflies and moths in sheer diversity. These flitting insects with scale-clad wings and sipping mouthparts coiled like a watch spring come in three times as many different kinds as all vertebrates combined. Almost all butterflies and moths pollinate some plant. Some, such as the famous yucca moth, reveal so intimate a partnership with vegetation of a few kinds that neither can reproduce sexually without the other. In alpine and arctic regions, where only an unusually warm and sunny day with little wind allows the scale-clad fliers to be airborne while flowers are open, the vegetation is notorious for propagating asexually. Years may go by without any benefit from insect pollen-carriers.

Naturalists in the teeming tropics notice a different aspect of the associations between local animals and plants. A surprisingly large number of different trees grow at some distance from the nearest neighbor of the same kind and depend upon the services of tropical fruit bats that unwittingly transfer pollen at night. Each bat lives for a dozen years or more, performing its various roles in the forest community. One after another flowering tree species offer free food as a lure to a bat. But if mankind finds some use for a particular bat-pollinated tree and removes every one that is large enough to produce a flower, what then? Unless the associated bats have a substitute, they can starve in a week. If they disappear from the community, fifty other types of trees that depend on bats may subsequently go untended and gradually vanish. George Herbert's story about the loss of a horseshoe nail and then shoe, horse, rider, battle, and kingdom (which Benjamin Franklin was so fond of quoting) has its real counterpart in the interdependence of many living things.

Most of the associates of any plant are like deer and rabbits in that they eat other kinds of vegetation. Or they hunt various small animals as prey. A truly limited diet is a liability that relatively few species can afford. Even many parasites, which are committed genetically to some special game plan in reaching the next appropriate host, can generally fulfill their destiny for a while at the expense of some substitute. That is why mankind suffers from infestations that properly belong to other hosts and wisely avoids the wild reservoirs or the transmitting agents. Malaria is a parasite of mosquitoes; typhus fever, a result of a rickettsia of rats; yellow fever, a virus of treetop monkeys; and psittacosis, a mild disease of birds. Yet each associate has an impact. It is an effective component in the environment. Along with insensitive vagaries, such as climate, it helps select those combinations of inheritance that can have a future.

Living things amaze us with their seemingly endless refinements in patterns of growth, of mature structure, of adaptive adjustments to environmental changes within the customary range. We accept that these refinements bought survival in a ruthless world and accumulated during an immensity of time. Today we notice the failures, the species becoming extinct. Their numbers increase at the same rate as the rise in human population. The inherited patterns are not good enough to cope in a world dominated and altered by our single species. At least an additional 10 percent of the earth's diversity in mammals, birds, cold-blooded vertebrates, and seed plants are threatened or so reduced in numbers as to be endangered and close to obliteration. These conservative measures of the present are documented in the "Red Data Books" issued and kept current in loose-leaf form by the prestigious International Union for the Conservation of Nature and Natural Resources, headquartered in Morges, Switzerland. Members of some national associations, such as the Audubon Society and the Institute of Wildlife Management express greater concern, estimating that 30 percent is a more realistic figure when the loss of original geographic range is combined with reductions in total numbers of well-known species.

A different criterion might be used. Are as many as 10 percent of native plants and animals in any country capable of associating continuously with mankind? These are the minority with a future while progressively more of the land and fresh water serve as productive resources for humanity. They are the ones that maintain a healthy population in neglected corners of a city, a suburb, an industrial park, a tree farm, or an agribusiness tract. Introduced species qualify too, such as the oriental tree of heaven (*Ailanthus*), which millions of people know as the "tree that grows in Brooklyn." Its roots spread under paved streets and sidewalks. Its trunk and foliage resist pollution by city dogs, automobile exhaust, incinerator fumes, and grime. In spring the tree expands its malodorous catkins to release pollen, and in autumn the fallen fruits and leaves add to the clutter below. Its green color in summer and its shade are welcome to people who see few alternatives. And the tree nourishes the large oriental silkworm, which was introduced separately; the silkworm now takes the place of the native cecropia, luna, polyphemus, and promethia caterpillars in producing one kind of large moth—a species that seems in no danger of disappearing no matter how run-down the neighborhood becomes.

These tolerant and opportunistic living things provide the principal remnant of a wild environment, interacting with each other and our species. It is they that may have the principal chance to rescue us from the egocentric situation of directing our own evolution. This responsibility would scarcely be undertaken deliberately, for it becomes awesome if we dream of a future extending for millions of years, rather than just the combined lifetimes of those personal descendants known to us.

Already we see disconcerting changes during a human lifespan. Just a few years ago we could talk with elders who vividly recalled great flocks of passenger pigeons or productive farmland where pavement and deteriorating buildings now await demolition and replacement. Over much of the earth, no children can experience the same streams, fields, and woods their parents knew at the same age, let alone those known to their grandparents. Nor, we feel certain, will the children of today's children comprehend the conversions and substitutions the world is going through at present. These changes follow no beneficial regularity that might allow us to regard them as a metamorphosis. They are not cyclic but progressive, leading rapidly in directions that have not been traveled, toward simplicities and fluctuations that will challenge the participants as never before.

Technologies that were developed in countries where literacy is almost universal allow us to travel easily into other parts of the world. We meet people where tribalism continues and few citizens see beyond a national boundary. The perceived values in life, human and otherwise, of the developing and the developed countries are in stark contrast. Often the leaders of newly independent nations lack the experience and perspective either to avoid the mistakes that have been made in older countries or to recognize how excellently in balance with local resources are the ways of their native people. Instead, the hope is to substitute foreign ways and quickly enjoy through imitation the coveted amenities of other lands. This course holds much greater attractions than the invention of a new life-style appropriate to the local situation. All too often the change requires both the elimination of familiar plants and animals by usurping their habitats and harsh measures to end tribal procedures that have for countless years harmonized with traditions and the environment. After tribal life has been disrupted and nonhuman life degraded, the facts emerge: the tools and fuels, the skills and markets that allow the transplanted ways to flourish elsewhere cannot be found in the new context. The majority of the native people, vegetation, and wildlife may be much worse off than before. Nor can the dream simply be abandoned and the former situation restored easily. Confidence in the old traditions has been shattered, and the native nonhuman life is largely gone. Yet ambitious people can rarely resist a gamble. They turn away from the advice a sagacious friend of ours offers upon occasion: "Never monkey with anything that's operating smoothly. It will be hard to get going again if it stops!"

Never before has the need been so great to turn human ingenuity toward keeping life running smoothly. In every community, some links in the natural system have already been lost. Introduced methods and alien species have not filled the gaps. Instead, too often, they have contributed to fluctuations and signs of instability. The world's ecology is admittedly out of joint. The present threat lies in pushing it so far that it cannot recover. The brink of catastrophe may lie closer than the horizon.

Epilogue

WHERE ARE WE?

We claim places on a small planet orbiting a star—our sun—of middle age and modest size. Fortuitously our world captures a small fraction of its energy and reradiates almost as much into space as useless heat. This balance holds the average temperature on earth within a range hospitable to life as we know it.

Life has been modifying its surrounding for a billion years or more. It continues to do so, in old ways and new. Green plants provide the oxygen in the atmosphere and renew it constantly. With oxygen, a sentient, swift-moving animal becomes a possibility. In the upper atmosphere, atoms of oxygen recombine in triplets, instead of pairs, and create a filter of ozone, which absorbs and converts to harmless heat the most dangerous component of sunlight. So long as the ozone is undiminished, this potent ultraviolet radiation cannot reach sea level or even a mountain top. Plants and animals can safely remain exposed to direct sunlight. The whole surface of earth stays open for colonization. The realm of life, which Lamarck called the *biosphere*, extends to the depths of the seas as well. Nothing like it has yet been found elsewhere, although other equivalents are probable on planets of distant stars.

About 90 percent of the kinds of living things inhabit the land and rely upon the fresh waters. Now mankind competes with all these plants and animals for space and moisture over almost all of this area. Our one species tries to manage fully 10 percent of the available fresh water, using about 120 gallons to produce each egg from a domestic hen, 300 to make a one-pound loaf of bread, more than 1,000 to get every pound of dried corn kernels, and 3,500 for each pound of beef. Already humanity has swept native plants and animals from all of the most productive land, where the climate is naturally favorable. Less suitable areas are being transformed too, still without raising a third of the human population much above a starvation level. Perhaps we should feel surprise that this competition has so far endangered only between 10 and 30 percent of the world's species.

Scientists fear the long-range changes that have been set in motion. At stratospheric heights the ozone filter is endangered by chemical reactions involving sunlight and waste gases from high-speed aircraft and pressurized spray cans. At lower levels in the biosphere, the fallout and by-products of man-made nuclear reactions add radioactivity that continues inexorably for thousands of years, stimulating mutations and cancers throughout the living world.

The early-warning systems in our modern life include plants and animals of no known commercial value. By noticing their condition, we can know in time to save ourselves, much as a coal-miner underground may do when a caged canary in the same tunnel shows distress. Lowly lichens, growing slowly on tree bark and tombstones, reveal where invisible gases (particularly sulfur dioxide) are poisoning the outdoor air. Many kinds thrive where air is pure, but few where the pollution continues. The few exposed rocks of Antarctica support more than four hundred different lichens, whereas Long Island, New York, has fewer than two dozen. Other plants and aquatic animals succumb in sequence to "acid rain," a phenomenon noticed first in Scandinavia, downwind from the factory chimneys of the English Midlands. Now the same acidified precipitation is evident in New England and elsewhere, as a dangerous corruption of the lower atmosphere from factory chimneys, paper mills, and electric generating stations that use coal or oil. Acid rain changes the nature of the soil, lessening its ability to retain the nutrients that decomposition makes available. Deprived, the vegetation shows malnourishment. The whole living community crumbles, unable to perpetuate itself. Ecologists and engineers hope to find a practical way to correct this before degradation affects too great an area.

The expedient answer sometimes proves to be merely a new problem. Air can be cleaned, its offending particles washed out and its noxious gases dissolved, by dirtying the water. The effects appear downstream instead of downwind. Aquatic plants and small animals, such as immature mayflies, become the indicators. They die, and then the fishes that could have eaten them as natural foods. Purifying the water, as a town or city must do to use it as a municipal supply, requires a new tradeoff, redistributing the costs and the nature of the wastes. Each step is a delaying action. It postpones the day when the cost of disposal will exceed the gains from production. Meanwhile, the ecological confrontation grows more acute.

The search for substitutes seems endless. If the whaling industry has overexploited the large marine mammals until most are verging on extinction, a new source of meat for pets must be found. Only one other product from whales—a liquid wax from the sperm whale—seemed irreplaceable. The automobile industry needs it for specialized lubricants in transmissions. Smaller quantities have a place in paper coatings, textile finishing, precision casting, polishes, and pharmaceuticals. Now a replacement has been discovered, extractable from the seeds of a desert shrub native to the American Southwest. A three-year study sponsored by the National Academy of Sciences led in 1975 to the promising conclusion that "the cultivation of jojoba, the manufacture of products from it, and the utilization of its by-products would greatly improve the economic situation of some native peoples and reservations in the area."

Now that whales are few, their feeding grounds in open seas must contain vast quantities of whale food, mostly krill crustaceans, resembling shrimp, two inches long or less. Inventors and investors who had profited in the past by providing efficient ships for pelagic whaling in southern oceans might devise and dispatch new equipment to harvest krill. The special craft would not have to resemble whales except in the ability to filter the crustaceans from surface waters. Guided by radio messages from scouts in helicopters, as pelagic whalers have been in recent years, the new equipment might find krill and transform it into products more useful to mankind than live whales. The operation should yield a desiccated crustacean meal, comparable to fish meal, valuable as a food supplement for livestock. The gains from harvesting and selling it might well repay the investment, including the costs of fossil fuel and human effort.

Every substitution for the benefit of mankind seems to affect wild plants or animals in an unfamiliar way. Sometimes the change does not exceed the ability of the nonhuman life to cope. We noticed this some years ago on Dassen Island, off the coast of South Africa, where guano-gatherers found seabirds—especially the delightful Cape penguins—contributing their nitrogenous wastes to a thick accumulation of hardened, sun-dried excrement. As rapidly as the human harvesters hauled away the guano, the birds adjusted their nesting habits. Instead of excavating burrows in the thick guano, the penguins dug obliquely into the softer sand that the harvesters exposed. The birds managed to raise their young, despite cave-ins that could bury eggs or hatchlings.

For a while, it seemed that the penguins were safe. Yet

Cape penguins on Dassen Island, near the southern tip of Africa, encounter many hazards because of mankind nearby: guano harvester, egg collectors, DDT and other pesticides in food, oil spills, and now outright competition for the fishes in cold offshore waters where fishermen come from far away. (Photo by Lorus J. and Margery Milne)

gradually their populations diminished to less than 10 percent. Fishermen were blamed for stealing penguin eggs; the government halted this activity. Then DDT was identified in unhatched eggs and dead chicks, although South Africa's coastal waters are linked only by ocean currents and winds to northern agricultural areas where this pesticide was used extensively. Gradually this menace too has faded. But penguins met a newer hazard: oil spills from tankers rounding the southern tip of Africa and outright competition for small fishes in the cold waters. Fouling with oil became serious right after Egypt closed the Suez Canal in 1967, forcing tanker captains to take the far more circuitous route. Reopening the canal in 1975 makes little difference, because supertankers have replaced the small vessels that could use the narrow man-made waterway.

The fish resource, which penguins and other seabirds have relied upon for millennia, presently attracts fishermen in ships of all sizes. Their nets leave behind far less food for those birds that dodge being caught. The great Benguela Current cannot supply enough for the fishermen whose ships are propelled by energy from oil and for the birds whose bodies depend on energy from fish.

Everywhere, the prodigal use of fossil fuels to subsidize civilization puts the ecology out of joint. Yet few people consistently aspire to the level of parsimony in use of energy that is universal in nonhuman life. It is as though an enlightened life-style for mankind could not be sustained without mortgaging the future and perhaps bankrupting the planet. No such dilemma confronted our ancestors when they turned in this direction almost three thousand years ago, at the first faint beginnings of the Iron Age. But the momentum developed in centuries past now carries us ever farther from the natural relationships among living things, even against our wiser judgment. It blocks us from taking seriously the quiet observation of Henry David Thoreau: "That man is richest whose pleasures are the cheapest."

The riches and frugality that Thoreau admired show everywhere that scientists explore into the untamed world. The riches of self-evolved, self-tested organization underlie the complex orchestration of each community, the subtle integration of the many members of every species, and the harmonious interaction among the inner parts of any individual. At the molecular level, we find this richness in the host of organic catalysts called for by the individual's inheritance. A separate gene specifies the formulation for each enzyme as life's key to a particular chemical reaction, easing it along in dilute solutions as a step in a promising pattern. Each step lets a living thing gain a little, while freeing a little energy in the form of heat. All of the steps combine into a significant sequence.

The sequence called forth by the inherited gene must

first be a private matter, important only to the individual. Yet in the long run, it is the diversity of inheritance within all individuals of a species that offers a continued future. "The fittest survive," we agree, and then compare the survivors with those that lost out to learn which fine details proved advantageous. We recognize that the perpetuation of the species surpasses in significance the prolongation of life in the individual. The ongoing success of the living community must be still more essential. The stability of the food web may be the loftiest goal of all, because it ensures the continued sharing of energy from sunlight through the many kinds and myriad individuals.

The world we share with more than 1.5 million other species is not ours, or theirs, to exploit. We belong to the living world and serve it. Each individual borrows chemical substances from the common (but limited) supply and remains indebted to every neighbor for life. That life depends upon a continued share of the energy as it flows through the food web from one kind of plant or animal to another. To maintain the flow, the community requires of its members a low-key conformity, each one solvent by living within its means and by affecting its neighbors under such exquisite control that they can stay solvent too. Any individual that ceases to contribute to the stability of the system becomes a liability, a mistake for natural selection to erase. Usually the erring one dies early. It pays off its debt to the environment by passing along its chemical components and some energy to neighbors with fitter ways.

The most inconspicuous weed or small creature perpetuates a significant heritage with a future that might be great. If it were not significant to the food web in some way, it would be a liability and soon extinct. Nor does history help us pick the winners in life to come, let alone single out mankind as the culmination. Who would have imagined that when trees began beckoning insects to proffered nectar and pollen, they would become the ancestors of two-thirds of the plant kingdom? Who would have expected their success to have helped downgrade the reptilian dinosaurs and let the descendants of six-inch, insect-hunting mammals inherit the lands and seas? Who could have anticipated that one of those descendants, a shuffling primate in East Africa a mere two million years ago, would have given rise to so many people in so many places, with so many ideas about the living world?

The idea that everything was created for human benefit, to be dominated and exploited, has not stood the test of time. The idea that mankind might coexist with the rest of creation by supporting the complex system instead of trying to simplify, control, or replace it has not yet had a chance. Yet we alone have evolved the capability to understand our indebtedness to nature, our dependence upon the welfare of the wild as well as of the tamed. Other kinds of life have made us what we are. But we, through our unique cultural evolution, have inherited a disproportionate responsibility for their tomorrows and ours as well.

One fundamental pattern, with helpful cycles within cycles, has served the living world since its beginning. Its recent breaks have put the ecology out of joint. Everyone can recognize the crisis situations and know that every life needs other life from beginning to end. Other life is the source and the continuation, not just of each species, but of all together.

Reading Around

DISCOVERIES MADE by using our senses and indirectly by sharing the ideas of others contribute to intellectual excitement. They give informed awareness a self-sustaining, self-propagating quality. Added depth and scope on some of the intriguing topics on which we touch can be explored in the sources of information and interpretation which we found especially worthwhile.

PROLOGUE

p. 3. The quotation from Elton comes from his fine *The Ecology of Invasions by Animals and Plants* (London: Methuen; New York: Wiley, 1958), pp. 111–12.

p. 7. The full wording of the act dated 1710 specifies trees "in Her Majesties Colonies of New-Hampshire, the Massachusets-Bay, and Province of Main, Rhode-Island, and Province-Plantation, the Narraganset Country, or Kings-Province, and Connecticut in New-England, and New-York, and New-Jersey, in America. . . ." A "keepsake" copy of the act was issued for private distribution by the editors of *Chronica Botanica*, Waltham, Massachusetts, September 5, 1949. We treasure one of these.

1. NATURAL DIVERSITY

p. 12. A South African ornithologist, D. Blaker, believes that cattle egrets have always been able to cross the South Atlantic Ocean. He points out in his article "Range Expansion of the Cattle Egret," *Ostrich,* Supplement 9 (1971), pp. 27–30, that before the increase of cattle ranching in the New World these birds associated with many different African grazers, whereas South America had only about fifteen species of large native herbivores (most of them forest-dwellers), offering any egrets that arrived virtually no opportunity to continue their inherited pattern of behavior. William J. Weber includes records of the rapid colonization of America in "New World for the Cattle Egret," *Natural History,* volume 81 (February 1972).

p. 13. The prospects for survival of the Asiatic lion, of which fewer than two hundred remain, are considered by Paul M. Tilden in "Last Stand of the Asiatic Lion," *National Parks and Conservation Magazine,* volume 44 (December 1970); and by Stephen Berwick, "The Gir Forest: An Endangered Ecosystem," *American Scientist,* volume 64 (January/February 1976).

p. 15. Frank N. Egerton, in "Changing Concepts of the Balance of Nature," *Quarterly Review of Biology,* volume 48 (June 1973), provided a thoughtful analysis of this concept and of the basic instability of living systems.

2. NEW ARRIVALS

p. 24. Ross's gull made the first page of the *New York Times* on March 4, 1975, the *Boston Globe* on the same day, *Time* magazine on March 17, *New Hampshire Audubon News* late in the month, and *The New Yorker* on April 7. Each account added extra facets to the story.

p. 32. Controversy over the coyote continues, with the hunters' and sheep ranchers' side well represented by articles in *Field and Stream, Outdoor Life,* and *Sports Illustrated,* plus one account, "Some Myths Concerning the Coyote as a Livestock Predator," by Maurice Shelton, in *BioScience,* this last evoking a storm of protesting letters in the June 1974 issue. We see more environmental concern and informative detail in the following:

Cole, John C. "The Return of the Coyote." *Harper's*, volume 246 (May 1973).
Gill, Don. "The Coyote and the Sequential Occupants of the Los Angeles Basin." *American Anthropologist*, volume 72 (August 1970).
McMahan, Pamela. "The Victorious Coyote." *Natural History*, volume 84 (January 1975).
Pringle, Laurence. "Each Antagonist in Coyote Debate Is Partly Correct." *Smithsonian*, volume 5 (March 1975).
Richens, Voit B. "Coyotes: Canids with a Challenge." *Maine Biologist*, volume 7 (January 1975).
Zeldin, Marvin. "Update: Coyotes and Lambs." *Audubon*, volume 77 (March 1975).

Older is our favorite *The Voice of the Coyote*, by J. Frank Dobie (Boston: Little, Brown, 1949), which retains its power to charm.

p. 37. The standard reference work *The Monarch Butterfly*, by F. A. Urquhart, was published in 1960 by the University of Toronto Press. Ongoing research is partly supported by the National Geographic Society, in the handsome magazine of which an illustrated report on new discoveries appeared in August 1976. Information on the heart poisons from milkweed in monarch caterpillars and adults appears at intervals in the pages of *Science*, in accounts by teams of researchers, such as those led by T. Reichenstein (August 30, 1968), R. I. Krieger (May 7, 1971), and Lincoln P. Brower (August 4, 1972).

3. CAST-OFF PETS

p. 44. Reporter Ed Buckow told the walking catfish story in *Field and Stream* (May 1969) somewhat after *Time* magazine offered the news "Fish Bites Dog" (August 23, 1968). A team of biologists at Florida Atlantic University, headed by Walter R. Courtenay, Jr., placed this addition in perspective in an article, "Exotic Fishes in Fresh and Brackish Waters of Florida," in *Biological Conservation*, volume 6 (October 1974). A broader analysis, commenting on the role of the pet industry, is by Courtenay with C. Richard Robins in *BioScience*, volume 25 (May 1975), entitled "Exotic Organisms: An Unsolved, Complex Problem"; it attracted letters from readers, published in the same journal in issues for October and December 1975.

p. 50. "Introduction of Tropical Fishes into a Hot Spring near Banff, Alberta," by D. E. McAllister of the Canadian National Museum of Natural Sciences, appeared in *Canadian Field Naturalist*, volume 83 (1969).

p. 55. Michel Desfayes, assistant to the secretary of the Smithsonian Institution, suggests that the collared turtle dove is essentially sedentary but shows an increase in range that matches the modification of the environment by mankind, particularly through cultivation. (Personal communication, October 1975)

p. 59. The monk parakeet in New York State was described in an illustrated article by Wayne Trim in the official publication *The Conservationist* for June/July 1972 and was later included in a new edition of *Birds of New York State* by John Bull (New York: Doubleday/Natural History Press, 1974). The number of different psittacine birds now loose in the United States can be appreciated more fully from the account by John Bull and Edward R. Ricciuti, "Polly Want an Apple?" *Audubon*, volume 76 (May 1974). The potential for new additions is evident from periodic reports such as "Birds Imported into the United States"; the one for 1972 was issued by the U.S. Fish and Wildlife Service as *Special Scientific Report—Wildlife*, No. 193, edited by Roger B. Clapp (1975).

p. 61. "Hounds of Hades" by Conway Robinson appeared in *Field and Stream*, volume 71 (December 1966). "Ecology of Feral Dogs in Alabama" is by M. Douglas Scott and Keith Causey of Auburn University, in *Journal of Wildlife Management*, volume 37 (July 1973). An editorial, "The Problem of Urban Dogs," in *Science*, volume 185 (September 13, 1974), attracted letters published in that journal in the issue for November 1. Soon afterward, *Time* magazine ran its cover story, "The Great American Animal Farm" in the issue for December 23, 1974.

p. 67. The introduction of the giant African snail into the Miami area of Florida was described by R. Tucker Abbott in *Nautilus*, volume 83 (October 1969); hasty efforts to exterminate the snails seemed successful by the end of 1972. The earlier history of this amazing mollusk is well described by Albert R. Mead of the University of Arizona in *The Giant African Snail: A Problem in Economic Malacology* (Chicago: University of Chicago Press, 1961).

p. 70. "Insects to Control Alligatorweed, an Invader of Aquatic Ecosystems in the United States," *BioScience*, volume 21 (October 1, 1971), by a research team headed by D. M. Maddox, presents a favorable view of this program. The chief officer of the Refuge Division in the Louisiana Wild Life and Fisheries Commission, Allan Ensminger, has protested that loss of alligatorweed due to beetle damage has promoted erosion, a decrease in the number of nutria as fur animals, and a reduction of marsh foods available to cattle. See *Catalyst,* volume 3 (1973).

p. 72. The category of "nuisance aquatic plants" is generating a long list of published studies that together reveal the ongoing controversy. The following have been especially informative:

Holm, L. G.; Weldon, W. W.; and Blackburn, R. D. "Aquatic Weeds." *Science,* volume 166 (November 7, 1969).

Sculthorpe, C. D. *The Biology of Aquatic Vascular Plants* (New York: St. Martin's Press, 1967).

Vietmeyer, Noel D. "The Beautiful Blue Devil." *Natural History,* volume 84 (November 1975).

Wilson, James S., and Boles, Robert J. *Aquatic Weeds* (New York: Stanford Printing Co., 1968).

U.S. Army Corps of Engineers. "Water-Hyacinth Obstructions in the Waters of the Gulf and South Atlantic States." *House Document* 37, 85th Cong., 1st Sess. (Washington, D.C.: Government Printing Office, 1957).

The possibility that the floating water fern *Azolla,* a common aquarium plant, may benefit the environment is raised by Arthur W. Galston in "The Water Fern–Rice Connection," *Natural History,* volume 84 (December 1975); and by P. J. Ashton and R. D. Walmsley in *Endeavour,* volume 35 (January 1976), in an article entitled "The Aquatic Fern *Azolla* and Its *Anabaena* Symbiont."

4. SOMETHING NEW TO EAT

p. 78. The discovery of the channeled whelk was reported by Rudolph Stohler in *"Busycotypus (B.) canaliculatus* in San Francisco Bay," *The Veliger,* volume 4 (1962). In the same journal, volume 11 (1969), James Carlton tried to explain the presence of *"Littorina littorea* in California (San Francisco and Trinidad Bays)"; and in volume 12 (1969), Richard L. Miller described his observations of the rockweed *"Ascophyllum nodosum* a Source of Exotic Invertebrates Introduced into West Coast Near-Shore Marine Waters."

p. 80. The oriental seaweed known locally as "Jap weed" (*Sargassum muticum*) was recognized as a potential pest by Louis D. Druehl in a letter on marine transplantations in *Science*, volume 179 (January 5, 1973), to which W. J. North responded (March 23, 1973). The arrival of the same seaweed and attempts to eradicate it are discussed by W. F. Farnham and E. B. G. Jones in *Biological Conservation*, volume 6 (January 1974), in their article "The Eradication of the Seaweed *Sargassum muticum* from Britain."

p. 81. Garald Boaich and H. T. Powel discussed "Introduction of Giant Kelp Proposed for European Waters" in *International Union for the Conservation of Nature Bulletin* (new series), volume 6 (February 1975).

p. 82. The spectacular spread of the Asiatic clam in North American waters can be traced in issue after issue of the journal *Nautilus*, beginning with volume 74 (January 1961). Sometimes the introduced species is called *Corbicula fluminea*, sometimes *C. manilensis*. An expression of alarm by W. M. Ingram, "Asiatic Clams as Potential Pests in California Water Supplies," *Journal of the American Water Works Association*, volume 51 (1959), now appears to be quite an understatement of the potential dispersal of this mollusk. David W. Behrens added an extra dimension to the story in *Nautilus*, volume 89 (January 1975), "The Use of Disposable Beverage Containers by the Freshwater Clam, *Corbicula.*"

p. 84. "Snails on Migrating Birds" are described by D. S. Dundee, P. H. Phillips, and J. D. Newson in *Nautilus*, volume 80 (January 1967).

p. 85. An optimistic review of carp history in America, with suggestions for better use of this fish, is "New Table Delicacy," by Frank DeLoughery, in *Environment*, volume 17 (March 1975).

p. 89. A representative of the Department of Fisheries at Jonkershoek, Cape Province, declared that South Africans first introduced *Tilapia mossambica* from Mozambique about forty years ago "to gobble up the algae" in fresh waters. So well did the fish clarify the streams and impoundments that many factories no longer needed chemical purifiers and some recreational waters became beautifully clear. The addition of fish cut costs and saved labor while producing good food. Afrikaaners call the fish the *korper*, whereas many English people in South Africa know it as the "mud bream." Officials of the Arizona Game and Fish Department introduced this species into drains near Yuma in the early

1960s. Carl Hubbs, an internationally respected authority on fishes, warned in *Transactions of the American Fisheries Society*, volume 97 (April 1968), that he feared another form of the "carp problem" if members of the family to which *Tilapia* belongs were introduced to America without extreme care; the benefits claimed all corresponded to the "good qualities of the carp" as promoted in 1873–75 in reports of the U.S. Fish Commission. Sure enough, two years later, *T. mossambica* was in California waters, as reported by Franklin G. Hoover and James A. St. Amant in *California Fish and Game*, volume 56 (January 1970).

p. 90. Anthony Netboy of Portland State College made himself the champion of Atlantic salmon. His *The Atlantic Salmon: A Vanishing Species* (Boston: Houghton Mifflin, 1968), like his article "The World's Most Harassed Fish" in *The Nation*, volume 207 (September 16, 1968), preceded fine accounts based upon his studies of Pacific salmon and their fight for survival against hazards introduced by technological man in *Natural History*, volume 78 (June/July 1969); in *American Forests*, volume 76 (March 1970); and in *Outdoor Life*, volume 151 (April 1973), the last a summary of the Danish-Canadian dispute over taking salmon on the high seas.

p. 93. John Laerum's success with raising Atlantic salmon in Norway became known through an article by Clive Gammon in *Sports Illustrated*, volume 35 (November 29, 1971).

p. 94. A good report on the fighting qualities of the kokanee salmon in eastern waters is "Little Salmon," by L. James Bashline, in *Field and Stream*, volume 76 (June 1971).

p. 95. The account of "Lampreys in the Lakes," by George F. Bush, in *Sea Frontiers*, volume 16 (May 1970), points out that the average adult lamprey in a lake makes 87 attacks on fish during its parasitic career, feeds 2,383 hours, and eats 18.6 pounds of fish, growing from 5.5 inches to 14.6 inches, between ages 5 and 7 years. The search for successful lampreycides is retold in "A Review of Literature on TFM (3-trifluormethyl-4-nitrophenol) as a Lamprey Larvicide," by Rosalie A. Schnick, in *Investigations in Fish Control*, number 44 (April 1972), U.S. Fish and Wildlife Service.

p. 96. An anonymous researcher for *Business Week* gave an account of the "Fish That Has Them Hooked: A Highly Successful Program of Breeding Coho Salmon for the Great Lakes," in number 2040 (October 5, 1968). Ted Janes extended the story in

"Cohos Come to New England," *Outdoor Life,* volume 149 (April 1972). As a footnote to praise for the fish ladder on the Lamprey River in Newmarket, New Hampshire, we might add that officers of the state fish-and-game department felt obliged to close the ladder at the peak of the 1975 fish migration because so many lampreys clung to the weirlike baffles that few migrants were getting through. Consequently the alewife crop in upper reaches of the river was the lowest since the ladder was built.

p. 100. The indigenous fishes of New Zealand suffered because of the introduction of trout and salmon from the Northern Hemisphere, as described by R. M. McDowall in *Transactions of the American Fisheries Society,* volume 97 (January 1968). But sports fishermen grow ecstatic over opportunities on both South Island and North Island, as is evident in separate accounts by Joe Brooks in *Outdoor Life,* volume 144 (December 1969) and 145 (January 1970).

p. 101. "The Introduction, Increase, and Crash of Reindeer on St. Matthew Island," by David R. Klein of the University of Alaska, is in *Journal of Wildlife Management,* volume 32 (April 1968). An account of attempts to raise reindeer on mainland Alaska, by Virginia Kraft, is in *Sports Illustrated,* volume 25 (December 12, 1966).

p. 105. Russel Kyle considered "Red Deer and the Future of Domestication," an alternative to sheep-raising, in *Animals,* volume 15 (April 1973). The potential suitability of the pronghorn is documented in David W. Kitchen's "Social Behavior and Ecology of the Pronghorn," *Wildlife Monographs,* number 38 (August 1974). Recommendations regarding "Big Game as Farm Livestock in Africa: Why Eland and Other Antelope Excel Cattle in Meat and Milk Production," by R. Whitlock, appear in *Field,* volume 233 (1969). The cultural obstacles were reviewed carefully in a symposium published as *Man, Culture, and Animals: The Role of Animals in Human Ecological Adjustments,* edited by Anthony Leeds and A. P. Vayda (Washington, D.C.: American Association for the Advancement of Science, 1965), AAAS Publication 78.

5. TARGETS AND TROPHIES

p. 113. *Introduced Mammals of New Zealand,* by K. A. Wodzicki (Wellington: Department of Scientific and Industrial Research,

1950), Bulletin 98, offers a long appendix citing dates of first successful introductions, classified according to basis, whether utility, feral, stowaway, or sport.

p. 114. The use of helicopter gunships to transform nuisance deer, tahr, and chamois into venison worth $1.60 per pound is described in "Slaughter on South Island" (New Zealand), by Robert F. Jones, in *Sports Illustrated,* volume 40 (March 18, 1974). It can be placed in perspective by reading "Eruption of Ungulate Populations, with Emphasis on Himalayan Tahr in New Zealand," by Graeme Caughley, in *Ecology,* volume 51 (1970); and "Behaviour Responses of Red Deer and Chamois to Cessation of Hunting," *New Zealand Journal of Science,* volume 14 (September 1971), by M. J. W. Douglas of the Forest Research Institute on South Island.

p. 117. The book *Pleistocene Extinctions: The Search for a Cause,* edited by Paul S. Martin and H. E. Wright, Jr. (New Haven: Yale University Press, 1967), has a fascinating sequel in "Simulated Overkill by Paleoindians," by James E. Mosimann and Paul S. Martin, in *American Scientist,* volume 63 (May/June 1975), plus letters and responses (July/August 1975); and in "Death of American Ground Sloths," by Austin Long and Paul S. Martin, in *Science,* volume 186 (November 15, 1974), followed by criticism and response in volume 191 (January 9, 1976). A similar view of man's role in "Breaking the Web" emerges in an analysis of recent extinctions in *Environment,* volume 16 (December 1974), by George Uetz and Donald L. Johnson.

p. 121. "Black Bear Predation on Salmon at Olsen Creek, Alaska," by George W. Frame, is in *Zeitschrift für Tierpsychologie,* volume 35 (August 1974).

p. 127. The history of ring-necked pheasants in North America is summarized well by sportswriter George Laycock in "Where Have All the Pheasants Gone?" *Field and Stream,* volume 74 (August 1969).

p. 128. "Say Goodbye to the New England Cottontail," by Michael Harris, in *New Hampshire Times,* volume 4 (April 2, 1975), quotes biologist Ted Walski of the state fish-and-game department, who said that "the loss of New Hampshire's farmland has left the animal with nowhere to live."

p. 130. The French study is reported by D. Ariagno and R. Delage, "Oiseaux et mammifères du Haut-Vercors," *Alauda,* volume 38 (1970).

p. 137. Aldo Leopold included a wealth of stimulating personal views in his *A Sand County Almanac* (New York: Oxford University Press, 1949), now available in paperback.

p. 139. "Conservation Is Not Enough" forms the final chapter in Krutch's *The Voice of the Desert: A Naturalist's Interpretation* (New York: William Sloan, 1954).

6. THE CLOCK MOVES ONLY ONWARD

p. 141. The adaptability of caribou might have been predicted from "Reaction of Reindeer to Obstructions and Disturbances" in Scandinavia, as described by David R. Klein in *Science,* volume 173 (July 30, 1971).

p. 143. In "Desert under the Trees," Malcolm Margolin discusses monoculture on tree farms that now total seventy million acres, an area equal to New York State and New England combined. Appearing in *National Parks and Conservation Magazine,* volume 44 (December 1970), the article quotes a report by the Tennessee Valley Authority predicting wildlife in the year 2000, in which wildlife management is defined as "the art and science of raising annual crops of wild animals for man's use."

p. 149. One of the earliest promotion schemes for mariculture, the "sea-cow fiasco" of 1893, is reviewed by Joseph J. Betz in *Sea Frontiers,* volume 14 (July/August 1968). The endangered status of these animals is central to Daniel S. Hartman's "Florida's Manatees—Mermaids in Peril," *National Geographic,* volume 136 (September 1969); and Bill Barada's "Manatees," *Animal Kingdom,* volume 74 (February 1971).

p. 153. Part of the ongoing story of the giant toad in new places can be gained from Taylor R. Alexander's "Observations on the Feeding Behavior of *Bufo marinus* (Linne)," *Herpetologica,* volume 20 (January 25, 1965); from Thomas Krakauer's "The Ecology of the Neotropical Toad *Bufo marinus* in South Florida," in the same journal, volume 24 (October 11, 1968); and from the news item on "The *Bufo* Plague" in Australia, *Time,* volume 104 (August 5, 1974).

p. 157. More on the Bermuda petrel is in our article "The Cahow—10 Years to Doom?" *Audubon,* volume 70 (Novem-

ber/December 1968). The cedar story is retold by David Challinor
and David B. Wingate in "The Struggle for Survival of the Bermuda Cedar," *Biological Conservation*, volume 3 (April 1971).

p. 162. Census tallies of Kirtland's warbler from 1931 to 1971
are discussed by L. H. Walkinshaw in *American Birds*, volume 26
(February 1972). Hazards that these birds meet are considered by
E. C. Mooney in *Defenders of Wildlife*, volume 49 (December 1973).

7. THE UNWANTED

p. 166. "The Mosquito: Still Man's Worst Enemy," says J. D.
Gillett of Brunel University, London, in *American Scientist*, volume
61 (July/August 1973).

p. 168. Anthony Wolff detailed the case history of the fire ant in
North America in "Big Schemes for Little Ants," *Audubon*, volume
73 (March 1971).

p. 173. At the end of February 1976, the manufacturer of Mirex
ceased production because the U.S. Department of Agriculture
discovered that the pesticide decomposed into Kepone when released on the land. Kepone is a banned substance because it
causes drastic nervous-system disorders in people. "A substitute is
now being sought to continue the fire ant program."

p. 174. In "The Brazilian Honeybee," written for *BioScience*,
published in volume 23 (September 1973), Charles D. Michener
reports the cautiously optimistic conclusions of a committee sent
to Brazil by the National Research Council in 1972. Roger A.
Morse, professor of apiculture at Cornell University, regarded the
dangers as grossly exaggerated when he visited Brazil in 1973, a
conclusion published in *Environment* (1975). These views are based
upon an expectation of change in the behavior of descendants of
African honeybees by the time they spread at their present rate to
North America—a hope with no guarantee.

p. 179. "Possible Effects of a Sea-Level Canal on the Marine
Ecology of the American Isthmian Region," by W. E. Martin,
J. A. Duke, S. G. Bloom, and J. T. McGinnis, in *Bulletin of the Battelle Memorial Institute*, number 171-44 (1970), indicates almost no
possibility for dispersal of sea snakes. William A. Dunson of Pennsylvania State University, who edited *The Biology of Sea Snakes* (Baltimore: University Park Press, 1975) and wrote the concluding

chapter, "Sea Snakes and the Sea Level Canal Controversy," would prefer to see a thermal barrier installed in any new canal to prevent sea snakes from getting through.

p. 184. A provisional list of "The 100 Most Dangerous Exotic Pests and Diseases" was compiled and distributed by the Plant Protection and Quarantine Program of the U.S. Department of Agriculture. It included thirty-five different insects and mites, none of them of sufficiently high economic impact to rank in the top sixteen, which included foot-and-mouth disease, African swine fever, fowl plague, and rinderpest.

8. GIVING TIME

p. 195. The amazing regeneration rate of kelp is evident in K. A. Clendenning's "Harvesting Effects on Canopy Invertebrates and on Kelp Plants," *Fisheries Bulletin of California,* number 139 (1967). Interactions between kelp and animals are analyzed by James A. Estes and John F. Palmisano in "Sea Otters: Their Role in Structuring Nearshore Communities," *Science,* volume 185 (September 20, 1974). Behavior is central to *The Sea Otter in the Eastern Pacific Ocean,* by Karl W. Kenyon, number 68 in the North American Fauna series published by the U.S. Fish and Wildlife Service (1969), now available in paperback from Dover Publications, New York. Kenyon also wrote an article, "Return of the Sea Otter," in *National Geographic,* volume 140 (October 1971). A fuller account of the kelp community is by James H. McLean in *Biological Bulletin,* volume 122 (February 1962).

p. 198. The toxin known as paralytic shellfish poison and its assay values in mollusks is discussed in " 'Red Tide' in the Southern Gulf of Maine, USA," by John Sasner, Jr., Miyoshi Ikawa, and Burdette E. Barrett, in *Biological Conservation,* volume 6 (January 1974), from an episode that peaked in mid-September 1972. Recurrences in several years are discussed by Wilson Wright in *Florida Naturalist,* volume 47 (December 1974). Success in "Predicting and Observing El Niño" is reported by an oceanographic team headed by Klaus Wyrtki of the University of Hawaii, in *Science,* volume 191 (January 30, 1976).

p. 202. Milner B. Schaefer's "Men, Birds and Anchovies in the Peru Current—Dynamic Interaction," *Transactions of the American*

Fisheries Society, volume 99 (1970), gives the basis for fuller under-standing of subsequent accounts, particularly the editorial "Anchovy Shortage Boosts Soybean Prices" in *Business Week*, number 2279 (May 12, 1973); then C. P. Idyll's view of the "Anchovy Crisis," *Scientific American*, volume 228 (June 1973); followed by the reported "Return of the Anchovy," *Newsweek*, volume 83 (May 6, 1974).

p. 209. The extinction of the great auk becomes more compre-hensible by reading Frederic A. Lucas's "The Expedition to Funk Island, with Observations upon the History and Anatomy of the Great Auk," *Report of the U.S. Nation Museum*, 1888–89 (1891); then E. Thomas Gilliard's "Bird Men Courageous," *Natural History*, volume 40 (September 1937); and his "The Bony Treasure of Funk Island," a chapter in *Discovery*, edited by John K. Terres (Philadelphia: Lippincott), pp. 46–58, sharing the thrill of explo-ration.

p. 210. The brown-and-white osprey, or fish hawk, is a world-wide indicator of pesticides in the environment. Reduced use of DDT has allowed the "Return of the Osprey," as Dennis Puleston entitled his article in *Natural History*, volume 84 (February 1975). Earlier a measurable "Decline of DDT Residues in Migratory Songbirds" had been reported by David W. Johnston in *Science*, volume 186 (November 29, 1974).

9. WHY FIGHT NATURE?

p. 223. In the midst of the discussions of blackbirds where they are not appreciated, the U.S. Fish and Wildlife Service published *Selected Bibliography on the Food Habits of North American Blackbirds*, Special Scientific Report—Wildlife, number 192 (1975). Although the impact of these feathered fliers when dispersed is generally beneficial to mankind, their winter concentrations are not. Frank Graham, Jr., asks if we can arrange to say, "Bye-bye Blackbirds?" *Audubon*, volume 73 (September 1971). Part of the problem is an introduced species, about which John W. Miller writes in "Much Ado about Starlings," *Natural History*, volume 84 (August/Sep-tember 1975).

p. 224. The practical side of the debate is in "Reducing Bird Damage to High Bush Blueberries with a Carbamate Repellant," by Charles P. Stone and others, in *Wildlife Society Bulletin,* volume 2 (Fall 1974). Contrasts between impatient management and natural control are discussed by William Murdoch and Joseph Connell of the University of California at Santa Barbara in "All about Ecology," published first in *The Center Magazine* (January 1970) and then reprinted in *Western Man and Environmental Ethics,* edited by Ian G. Barbour (Reading, Mass.: Addison-Wesley, 1973). The complex involvement of fruit bushes, pulpwood, birds, insects, and pesticides is brought out by Jeanne Huber in "The Blueberry Birds of Brunswick," *Sierra Club Bulletin,* volume 59 (March 1974).

p. 226. Faith McNulty's book *Must They Die? The Strange Case of the Prairie Dog and the Black-Footed Ferret* (New York: Doubleday, 1971) was generously condensed in *The New Yorker* (June 13, 1970) and summarized in "The Black-Footed Ferret," *National Parks and Conservation Magazine,* volume 45 (May 1971). Much more on the interactions can be gleaned from *Proceedings of the Black-Footed Ferret and Prairie Dog Workshop* held in 1973 in Rapid City, South Dakota, edited by R. L. Linder and C. N. Hillman (Brookings: South Dakota State University, 1974).

p. 231. Thomas R. Vale analyzed "Sagebrush Conversion Projects: An Element of Contemporary Environmental Change in the Western U.S." in *Biological Conservation,* volume 6 (October 1974).

p. 235. Early conflicts over "The Porpoise and the Tuna" are described by William F. Perrin in *Sea Frontiers,* volume 14 (May/June 1968). Later alternatives are evident in "The Dolphin-Tuna Controversy," by Stanley M. Minasian, in *Pacific Discovery,* volume 28 (January/February 1975).

p. 238. Sportswriter George Laycock wrote "The Troubled Alligator," *Field and Stream,* volume 71 (April 1967); and John C. Ogden, "Survival of the American Crocodile in Florida," *Animal Kingdom,* volume 74 (December 1971). The American crocodile was declared in 1975 the "rarest reptile in the U.S." In Africa, a start has been made on "Crocodile Rearing in Zululand," according to Tony Pooley, in *Animals,* volume 13 (June 1970).

p. 240. Welcome news of "The Thames—A River Reborn" is by Peter Grant and Jeffrey Harrison, announced in *Animals,* volume 15 (May 1973), with census details showing the numbers of mallard, pintail, and pochard up, of widgeon, down.

p. 244. The role of guppies is discussed in "Precarious Odyssey of an Unconquered Parasite," by Kenneth S. Warren, in *Natural History*, volume 83 (May 1974).

p. 245. Mosquitos infected with bird malaria and bird pox are indicted as the introduced agents serving "To Kill a Honeycreeper," discussed by Richard E. Warner in *Natural History*, volume 82 (August/September 1973).

10. UNEQUAL HORIZONS

p. 256. In *National Parks Magazine*, volume 44 (October 1970), S. Blackwell Duncan reviews briefly the history of human persecution of the wolf. *The Wolf: The Ecology and Behavior of an Endangered Species*, by L. David Mech (Garden City: Natural History Press, 1970), is full of details, including some from the doctoral dissertation of Mech's student Philip C. Shelton, "Ecological Studies of Beavers, Wolves, and Moose in Isle Royale National Park, Michigan" (1966), summarized in *Dissertation Abstracts*, volume 27 (7B). Two sides of the controversy are presented by sportswriter Ben East in "Who Are His Real Friends? Battle over the Timber Wolf," *Outdoor Life*, volume 152 (September 1973); and by Nathaniel Reed, assistant secretary of the Interior Department, "Wolf Controversy," in the same magazine, volume 153 (March 1974). Mech provided a "New Profile for the Wolf," *Natural History*, volume 83 (April 1974); and Durward L. Allen (Mech's former professor), the article "Of Fire, Moose, and Wolves," *Audubon*, volume 76 (November 1974).

p. 257. The natural role and present status of *The Puma: Mysterious American Cat* is easier to comprehend now that Stanley P. Young's 1946 book has been reissued in paperback by Dover Publications (1964); Maurice G. Hornocker's "The American Lion" appeared in *Natural History*, volume 79 (November 1970); and Randall L. Eaton reported on the surviving "Florida Panther" in *National Parks and Conservation Magazine*, volume 45 (December 1971).

p. 263. The poetic view is expressed in "In Praise of Prairie" among *The Collected Poems of Theodore Roethke* (Garden City: Doubleday, 1975).

p. 264. James W. Tilden of San Jose State College wrote "The

Insect Associates of *Baccharis pilularis* DeCandolle," *Microentomology*, volume 16 (April 1951).

p. 266. We enjoy associating L. Hugh Newman's article "When Churchill Brought Butterflies to Chartwell," Kent, *Audubon*, volume 67 (May/June 1965), with Paul Showers's "Signals from the Butterfly," *New York Times Magazine* (July 27, 1975).

p. 269. Cynthia, the "Silk Moth of the Railroad Yards," was introduced in an article by Robert M. Pyle with photographs in *Natural History*, volume 84 (May 1975).

EPILOGUE

p. 272. In March and April 1975, the *Federal Register* from Washington, D.C., carried short notices from the Department of the Interior formally requesting the views of governors and other interested parties on the status of specific butterflies and vertebrate animals for which the Fish and Wildlife Service had evidence suggesting a need for protection. On July 1, 1975, the *Federal Register*, devoted 101 pages to threatened, endangered, or possibly extinct plants for which similar conservation efforts seemed necessary.

p. 272. "Lichens as Indicators of Pollution," by Irwin M. Brodo of the National Museum of Canada, makes a disturbing illustrated article in *The Conservationist* for August/September 1971. Acid rain, with a deleterious effect on forests and fisheries, is particularly serious in northern areas where the thin soil that has accumulated since the Ice Age has little buffering capacity; decomposition under coniferous trees enhances the acidification. The relationship between "Acid Precipitation and Embryonic Mortality of Spotted Salamanders, *Ambystoma maculatum*," as reported by F. Harvey Pough in *Science*, volume 192 (April 2, 1976), points to further deterioration of the environment.

p. 273. Milton Harris served as chairman of the Committee on Jojoba Utilization for the National Academy of Sciences, from which a report, *Products from Jojoba: A Promising New Crop for Arid Lands*, became available in 1975.

p. 274. "Whales and Krill in the Twentieth Century," by N. A. Mackintosh, is a chapter in *Antarctic Ecology*, edited by M. W. Holgate, volume 1 (New York: Academic Press, 1970).

Our account of the Cape penguins of South Africa, the numbers of which have shrunk to less than 4 percent of the former population, appeared as "The Last Survivors" in *International Wildlife*, volume 5 (May/June 1975).

p. 276. Thoreau penned this comment in his *Journal* on March 11, 1876. Will his view be more widely held when the Tricentennial of the United States occurs a century hence?

p. 277. We share in the growing anxiety over actions that keep the ecology out of joint. A quick move, perhaps as unthinking as tossing a pebble into a pool, sends ripples in so many directions. "Did a Barbados Hunter Shoot the Last Eskimo Curlew?" asks Mary W. Bond in *Audubon*, volume 67 (September/October 1965), referring to a bird shot at sunset on September 4, 1963, given to Captain Maurice B. Hutt, professor of history at Harrison College, Barbados. He put it in the deep-freeze chest and later gave its skin to the Philadelphia Academy of Sciences. "Extinction Strikes 10,000 Species; Man Remains Unconcerned," writes A. Solem, in *Bulletin of the Field Museum of Natural History*, volume 41 (1970). It is not enough for biologists to read "Preservation of Natural Diversity: The Problem of Extinction Prone Species," by John Terborgh of Princeton University, in *BioScience*, volume 24 (December 1974); the need is for everyone to strive together to save what is left of the living world before time runs out.

Index